New Code of

Estimating Practice

New Code of

Estimating Practice

The Chartered Institute of Building

Roger Flanagan
Carol Jewell

WILEY Blackwell

This edition first published 2018

Registered Offices
John Wiley & Sons, Inc., 111 River Street, Hoboken, NJ 07030, USA
John Wiley & Sons Ltd, The Atrium, Southern Gate, Chichester, West Sussex, PO19 8SQ, UK

Editorial Office
9600 Garsington Road, Oxford, OX4 2DQ, UK

For details of our global editorial offices, customer services, and more information about Wiley products visit us at www.wiley.com.

Wiley also publishes its books in a variety of electronic formats and by print-on-demand. Some content that appears in standard print versions of this book may not be available in other formats.

Library of Congress Cataloging-in-Publication Data applied for
Paperback ISBN: 9781119329466

Cover design by Wiley
Cover image: © Zephyr18/Gettyimages

Set in 10/13pt ITCFranklinGothic by SPi Global, Chennai, India

10 9 8 7 6 5 4 3 2 1

Contents

Foreword

Developments in technology and the globalisation of supply chains have changed our industry in ways that were unimaginable a generation ago, and they continue to shape our industry and challenge us to find new ways of working.

The modern construction project is often a complex undertaking that depends on a highly skilled international workforce, access to products and services from around the world and the expertise to co-ordinate and manage a process that is, by its nature, risky.

For a construction company, the consequences of getting it wrong can be catastrophic. Now, more than ever, there is a need to understand why projects go wrong and to act to ensure that lessons learned are applied to future projects.

Understanding how the price and performance risk inherent in construction projects is managed is key, and this means looking more closely at the process of estimating and tendering.

This *New CIOB Code of Estimating and Tendering Practice* examines that process and provides best practice guidelines for those involved in procuring and pricing tenders for construction works both in the public and private sectors. The principles that it describes are applicable to any project regardless of size or complexity.

I commend this publication as part of the CIOB's ongoing commitment to raising standards of professionalism in our industry.

Paul Nash MSc PPCIOB
Past President 2016-17

Glossary

This glossary gives explanations of the terms used in this book and in the general field of construction cost estimating.

All-in labour rate a rate which includes payments to operatives and associated costs which arise directly from the employment of labour.

All-in material rate a rate which includes the cost of material delivered to site, waste, unloading, handling, storage and preparing for use.

All-in mechanical plant rate a rate which includes the costs originating from the ownership or hire of plant, together with the operating costs of fuel, driver and insurance.

Approved contractors those who have demonstrated that they have the expertise, resources, ability and desire to tender for a proposed project. Selection of such contractors is normally by pre-selection procedures.

Attendance the labour, plant, materials and/or other facilities provided by main contractors for the benefit of sub-contractors/specialty contractors, for which they normally bear no cost.

Awarding authority the public sector body (department agency, NHS Trust, local authority etc.), which is procuring a service through the private finance initiative (PFI).

Base cost estimate an estimate of the cost of the project without cost risk, uncertainty and inflation. The basis on which the base cost estimate has been prepared needs to be recorded and understood.

Base date of cost data the date on which rates and prices contained within a cost analysis or benchmark analyses are taken as the base date for calculations.

Benchmarking a procedure for testing whether the standard and price of services is consistent with the market standard, without any formal competitive tendering. This is usually adopted during the project concession period to ensure facility management services continue to represent value for money.

Bid bond a written form of security executed by the bidder as principal and by a surety for guaranteeing that the bidder will sign the contract if awarded the contract for the stated bid amount.

Bid rigging this arises when some or all of the bidders in a competitive tender attempt to frustrate the purpose of the competitive tender by limiting the degree of competition in it. Bid rigging is illegal and considered to be a form of corruption.

Bill of quantities (BoQ) a document prepared by the cost consultant to provide project-specific measured quantities of the items of work identified by the drawings and specifications in the tender documentation. The BoQ forms part of the contract documents in many countries.

BIM (Building Information Modelling) the process of generating and managing data about the building during its lifecycle. Typically, BIM uses three-dimensional, real-time, dynamic building modelling software to increase productivity in the design and construction stages.

Buildability the extent to which the design of a building facilitates ease of construction subject to the overall requirements of a completed building.

CDM regulations *Construction (Design and Management) Regulations 2015* concern the management of health and safety. These regulations impose duties on clients, planning supervisors, designers and contractors.

Component a small self-contained part of a building, constructed from a number of smaller parts, and designed to perform a particular function, for example, a window or a pump. It is a measured item that forms part of an element or a sub-element which is cost estimated to ascertain the cost target for an element or a sub-element.

Construction inflation an upward movement in the average level of prices and/or costs. An allowance for any fluctuations in labour, plant and equipment and materials prices needs to be made in the cost estimate/plan. See Tender inflation.

Consultants the client's or contractor's advisers on design, engineering, cost and other matters. Such advisers may include project managers, architects, engineers, quantity surveyors, accountants, bankers or other persons having expertise of specific areas.

Contingency sum an undefined provisional sum of money required by the employer to be included in the tender sum for unknown work.

Contingency period a period of time allocated in the tender works programme for contractor's time-risk events.

Contractor's estimate net estimated cost of carrying out the works for submission to management at the final review meeting.

Cost the estimated cost of the physical production of work. (Note: Estimated cost should not be confused with historical cost; historical cost is the cost of construction which is revealed only after the work has been executed.)

Cost breakdown structure (CBS) the separation of items, resources, tasks and projects and their associated costs, cost targets and values. The ability to separate costs provides greater financial control. Comparisons of estimated against actual cost for individual items ensure that those items that result in loss and those that make a profit can be identified.

Cost checks (cost check or cost checking) this is a method of controlling the cost of a project within a pre-determined sum, during the design process. It includes the preparation of the cost plan and the subsequent stages of cost checking.

Cost control the monitoring of costs for comparison with the project budget in order that control decisions can be taken.

Cost limit (or authorised budget or approved estimate) the maximum expenditure that the employer is prepared to make in relation to the completed building.

Cost per functional unit (or functional unit cost) the unit rate which, when multiplied by the number of functional units, gives the total building works estimate (i.e. works cost estimate less main contractor's preliminaries and main contractor's overheads and profit). The total recommended cost limit (i.e. cost limit, including inflation) can also be expressed as a cost per functional unit when reporting costs.

Cost per m² of gross internal floor area (or cost/m² of GIFA) the unit rate which, when multiplied by the gross internal floor area (GIFA), gives the total building works estimate (i.e. works cost estimate less main contractor's preliminaries and main contractor's overheads and profit). Other cost estimates that form part of an order of cost estimate or a cost plan, should also be converted to costs/m² of GIFA when reporting costs to the employer and project team (i.e. to express cost targets for group elements, elements, sub-elements, as well as the cost limit). They are also used in cost analyses and benchmarking as a means of documenting costs of previously completed building projects.

Cost plan the report or budget prepared by a Quantity Surveyor for the purpose of establishing managing the budget for the elemental categories.

Cost records records of historical costs and notes of the conditions prevailing when such costs were incurred.

Cost target a pre-estimate of the most likely outturn cost for the project as defined in the contract documentation. In a bill of quantities, it is the total expenditure for an element or work package.

Credit for materials a refund offered by the contractor to the employer in return for the benefit of taking ownership of materials, goods, items, mechanical and electrical plant and equipment, etc. arising from demolition or strip-out works.

Critical path method a technique used to predict project duration by analysing which sequence of activities has the least amount of scheduling flexibility. Early dates are figured by a forward pass using a specific start date, while late dates are figured by using a backward pass starting from a completion date.

Daywork payment agreement to undertake work for the price of the labour and materials, with a percentage to cover overheads and profit. The method is generally used for unforeseen events or variations where the work was not specified in detail within the contract or were not covered due to the circumstances. Where work is undertaken using this method, the daywork sheets, describing the work done, material used, and the labour, are signed off at the end of the day, or period of work, by the resident engineer, clerk of works or client representative.

Defined provisional sum for a defined provisional sum, contractors must be given full information about the nature and extent of the work. They are required to make provision in their tender works programme for an adequate duration and sequence for the work together with associated preliminaries. See Provisional sums.

Deflation a downward movement in the average level of prices and/or costs (i.e. the opposite of deflation). It is included as an allowance in the order of cost estimate or cost plan for fluctuations in the basic prices of labour, plant and equipment and materials. See Tender inflation and Construction inflation.

Design and build contracts where project documents are compiled with the contractor's design obligations relating to the whole of the works in mind.

Design Management the management of the process of design. It is the responsibility of the design manager to ensure that necessary framework is in place to allow the design team to produce fully coordinated and complete information in a timely manner. Design management can be undertaken by the design team or the construction team.

Design team architects, engineers and technology specialists responsible for the conceptual design aspects of a building, structure or facility and their development into drawings, specifications and instructions required for construction and associated processes. The design team is a part of the project team.

Director's adjustment a reduction or addition to the tender price, derived by the contractor's estimating team, judged by the director(s) of the contractor.

Domestic sub-contractors sub-contractors selected and employed by a contractor.

Down time (or standing time) the period of time that the plant is not operating. This may be due to breakdown, servicing time or an inability to operate due to other factors.

E-procurement a public procurement procedure initiated, conducted and/or concluded using electronic means, that is, using electronic equipment for the processing and storage of data, in particular through the Internet.

E-tendering an electronic system, described as a set of functionalities, to support the tendering processes.

Effective rate the price rate calculated by dividing a gang cost by the number of productive operatives in the gang.

Element buildings are divided into elements (e.g. foundations, external envelope, roofs etc.). An element is that part of any building that always performs the same function irrespective of its construction or specification. A separate cost target can be established for each element.

Element unit quantity (EUQ) the unit of measurement relates solely to the quantity of the element or sub-element (e.g. the area of the external walls, the area of windows and external doors and the number of internal doors).

Element unit rate (EUR) calculated by dividing the total cost of an element by the element unit quantity (EUQ). EURs include the cost of all materials, labour, plant, subcontractor's preliminaries, subcontractor's design fees and subcontractor's overheads and profit. They exclude main contractor's preliminaries, main contractor's overheads and profit and other allowances, such as project/design team fees, other development/project costs, risk allowances and inflation. These items should be assessed separately.

Elemental cost analysis (or cost analysis) a full appraisal of costs of previously constructed buildings. The aim of the analysis is to provide reliable information to assist in the accurate estimation of costs for future buildings. It provides a product-based cost model, providing data on which initial elemental estimates and elemental cost plans can be based.

Elemental cost plan (or cost plan) the critical breakdown of the cost limit for the building(s) into cost targets for each element of the building(s). It provides a statement of how the design team proposes to distribute the available budget among the elements of the building, and a frame of reference from which to develop the design and maintain cost control. It also provides both a work breakdown structure (WBS) and a cost breakdown structure (CBS), which, by codifying, can be used to redistribute work in elements to construction works packages for the purpose of procurement.

Elemental method a budget setting technique which considers the major elements of a building and provides an order of cost estimate based on an elemental breakdown of a building project. The elemental method can also be used to develop an initial cost model as a prerequisite to developing an elemental cost plan. The method involves the use of element unit quantities (EUQ) and element unit rates (EUR).

Employer the owner and/or the developer of the building; in some cases the ultimate user. The terms senior responsible owner (SRO) and project sponsor are used by central civil government and the defence sector; being the representatives empowered to manage the building project and make project-specific decisions.

Enabling works a generic description for site preparation works that might take place prior to work under the main construction contract. For example:

Demolition, Site clearance, Tree protection, Diversion and/or disconnection of existing site services, Geotechnical and exploratory ground investigation, Decoupling from existing buildings and Decontamination.

Estimating the technical process of predicting the costs of construction.

Estimator a person performing the estimating function in a construction organisation. Such a person may be a specialist or he/she may carry out the estimating function in conjunction with other functions, such as quantity surveying, buying, planning or general management.

Estimate base date the date on which the cost limit is established as a basis for calculating inflation, changes or other related variances.

Facilitating works specialist work that needs to be completed before any construction of the facility can begin; for example, removal of hazardous material.

Facilitating works estimate the sum of the cost targets within the 'Facilitating works' group element. It excludes the building works estimate, as well as those related to main contractor's preliminaries, main contractor's overheads, profit/design team fees estimate, other development/project costs estimate and risk allowances. See Group element.

Final review the action taken by management to convert an estimate into a tender. Also commonly known as 'appraisal', 'settlement' or 'adjudication'.

Firm price contract a contract where the price is agreed and fixed before construction starts.

Fixed charge charge for work, the cost of which is to be considered independent of duration.

Fixed price contract a contract where the price is agreed and fixed before construction starts. The term 'firm price' is used to denote more precisely a contract which will not be subject to fluctuations.

Fluctuations the increase or decrease in cost of labour, plant, materials and/or overhead costs which may occur during a contract.

Formal cost plan the elemental cost plan which is reported to the employer on completion of a specific Work Stage (RIBA) or OGC Gateway.

Formal cost plan stage the point at which the quantity surveyor/cost manager formally submits an elemental cost plan to the employer for consideration. The formal cost plan stages are interlinked with the appropriate RIBA Work Stages and OGC Gateways.

Functional unit a unit of measurement used to represent the prime use of a building or part of a building (e.g. per bed space, per house and per m^2 of retail area). It also includes all associated circulation space.

Functional unit method a budget-setting technique which consists of selecting a suitable standard functional unit of use for the project, and multiplying the projected number of units by an appropriate cost per functional unit, such as cost per bed space in a hospital.

Gang cost a grouping of labour costs to include principal and supporting labour associated with a particular trade. It may also include items of plant.

General plant part of a contractor's project overhead calculation for plant excluded from unit rate calculations and which is available as a general facility on site. Durations for general plant are usually taken from the tender programme.

Gross external area (GEA) the sum of the floor areas contained within the building measured to the external face of the external walls, at each floor level.

Gross internal floor area (GIFA) (or gross internal area (GIA)) the floor area contained within the building measured to the internal face of the external walls at each floor level.

Group element the main headings used to describe the facets of an elemental cost plan (i.e. Substructure; Superstructure; Internal finishes; Fittings, furnishings and equipment; Services; Complete buildings and building units; Work to existing buildings; External works; Facilitating works; Main contractor's preliminaries; Main contractor's overheads and profit; Project/design team fees; Other development/project costs; Risks; and Inflation).

Head office overheads the cost of administering a company and providing off-site services. The apportionment of head office overheads, to individual projects or as a percentage of company turnover, is decided by management as part of management policy.

JCT The Joint Contacts Tribunal, responsible for producing the standard forms of building contract.

Labour-only sub-contractors sub-contractors whose services are limited to the provision of labour.

Lump sum contract a fixed price contract where contractors undertake to be responsible for executing the complete contract work for a stated total sum of money.

Main contractor (or prime/principal contractor) the party who has the main contract with the client is responsible for the work on site and the employment of specialty/subcontractors and specialists necessary to complete the construction work. Some contractual arrangements may result in contractors being employed directly by the client; in such cases, the main contractor would be the party responsible for the majority of the construction work.

Main contractor's overheads the main contractor's fixed costs associated with running the company including salaries, operating expenses etc., proportioned to each building contract.

Main contractor's preliminaries the cost to the main contractor of running a site. These include the costs associated with management and staff, site establishment, temporary services, security, safety and environmental protection, control and protection, common user mechanical plant, common user temporary works, the maintenance of site records, completion and post-completion requirements, cleaning, fees and charges, sites services and insurances, bonds, guarantees and warranties. Main contractors' preliminaries exclude costs associated with subcontractors' or work package contractors' preliminaries.

Management those responsible for the function of general management and having the responsibility for making the decision to tender and for reviewing tenders.

Market testing a procedure for re-pricing the provision of services on a periodic basis by means of a competitive tender.

Mark-up the sum added to a cost estimate, following the final review meeting, to arrive at a tender sum. Mark-up will include margin, allowances for exceptional risks, and adjustments for commercial matters such as financing charges, cash flow, opportunities (scope) and competition. There may be a requirement for main contractor's discount when tendering as a sub-contractor, and value-added tax when required in the tender instructions.

Master construction programme the name given to the contractor's overall programme for the works under the JCT form of contract. The programme should contain no more than 2,000 activities of durations no greater than 1.5 times the progress reporting period, or it will be too cumbersome to maintain. It should be a critical path network and show, among other things, all contract requirements of dates for possession and completion, the sequence of planned durations, the logic of the principal activities and the critical path (or paths) to every completion date.

Method statement a statement of the construction methods and resources to be employed in executing construction work. This statement is normally closely linked to a tender programme.

Net internal area (NIA) the usable area within a building measured to the internal face of the perimeter walls at each floor level.

New Rules of Measurement (NRM) published by the Royal Institution of Chartered Surveyors (RICS), they provide a standard set of measurement rules for estimating, cost planning, procurement and whole-life costing for construction projects.

Nominated sub-contractor/supplier a sub-contractor/supplier whose final selection and approval is made by the client or client's advisers. See Prime cost.

Off-site Pre-fabrication the use of a specially designed manufacturing facility to construct building modules that are delivered directly to the site. Design for manufacture and assembly (DfMA) is specified more frequently. Design for manufacture and assembly includes the use of prefabrication and off-site manufacture, which includes modular or volumetric units, flat-pack, or panel systems using standard components. The key principle is to leverage benefits by using standardised, repeatable processes and designs. Standardisation also leads to rationalisation, and optimisation.

Open competitive tendering an impartial method of procurement whereby contractors are invited through advertisements to apply for tender documents. The number of tenderers is not usually limited, and reputation and ability to execute the work satisfactorily are not always considered.

Option cost an estimate of the cost of alternative design solutions to achieve the employer's objectives, so that they can be compared and appraised. Option costs will be incorporated in the overarching cost report.

Order of cost estimate the determination of possible cost of a building(s) early in design stage in relation to the employer's fundamental requirements. This takes place prior to preparation of a full set of working drawings or bills of quantities and forms the initial build-up to the cost planning process.

Other development/project costs costs that are not necessarily directly associated with the cost of constructing the building, but form part of the total cost of the building project to the employer (e.g. land acquisition costs, fees for letting agents, marketing costs and contributions associated with planning permissions).

Output specification the specification that sets out the requirements in non-prescriptive terms, so that the tenderers can determine how to provide the services.

Overheads and profit see Main contractor's overheads and profit.

PDM (Precedence diagram method) a method of constructing a logic network using nodes to represent the activities and connecting them by lines that show dependencies.

PFI (Private finance initiative) initially developed by the United Kingdom to provide financial support for public–private partnerships between the public and private sectors. It has now been adopted throughout the world as part of a wider programme for privatisation and deregulation driven by corporations and governments and international bodies.

Post-tender estimate prepared after all the construction tenders have been received and evaluated. It is based on the outcome of any post-tender negotiations, including the resolution of any tender qualifications and tender price adjustments. The post-tender estimate will include the actual known construction costs and any residual risks.

Preliminaries the costs of running a site as a whole rather than any particular zone or any particular activities. They are sometimes referred to as site overheads (or in the United States as 'field costs'). Thus, it is essential that each case is inspected individually to determine which resources are affected by a delay or disruption irrespective of how the contractor has priced the resource. See Main contractor's preliminaries.

Pre-qualification the provision by a contractor of information as part of a pre-selection process. An application by a contractor to be included on a select list of tenderers.

Pre-selection the establishment of a list of contractors with suitable experience, resources, ability and desire to execute a project, bearing in mind the character, size, location and timing of a project.

Pre-tender estimate a cost estimate prepared immediately before calling tenders for construction.

Prime cost sum (PC sum) the amount included in a contract for work that is foreseen but cannot be accurately specified at the time the tender documents are issued.

Procurement the process which creates, manages and fulfils contracts related to the provision of supplies, services or engineering and construction works, the hiring of anything, disposals and the acquisition or granting of any rights and concessions.

Profit the amount of money that remains after all project and company expenses have been paid. This can be assessed either as a return on the investment or as compensation for the risk assumed by the construction company owners.

Project team employer, project manager, quantity surveyor/cost manager, design team and all other consultants responsible for the delivery of the building project on time, on cost and to the required performance criteria (design and quality). The project team will include the main contractor where the main contractor has been engaged by the employer to provide pre-construction services.

Project/design team fee(s) the cost of design, including consultants' fees and contractors' design and pre-construction fees, for example, management and staff, specialist support services, temporary accommodation services and facilities charges and the main contractor's overheads and profits.

Project/design team fees estimate the total estimated cost of all project/design team fees at the estimate base date (i.e. excluding tender inflation and construction inflation).

Project overheads (sometimes referred to as site overheads, general cost items or expenses). The cost of these site-specific project costs that cannot be allocated to individual activities and that are not included in all-in or composite rates. Amongst other things, these costs may include site management, huts, safety precautions, job-related insurances, bonding costs, telephone, water, electricity costs etc. The essential characteristic is that these overheads serve more than one activity, for example, tower cranes, skips, general site labour etc. However, in practice, some resources that could be allocated to an activity, for example, scaffolding for falsework, are included in the preliminaries because of the contractor's preferred method of pricing. See also Main contractor's preliminaries and Preliminaries.

Provisional quantity a quantity included in the Bill of Quantities for works whose nature and scope cannot be entirely foreseen or defined at the time of tendering.

Provisional sums the New Rules of Measurement for Building Works (NRM) provides for sums that may be included in tender documents for work which cannot be measured at the tender stage. These sums are inclusive of overheads and profit allowances. There are two types of provisional sum: 'defined' and 'undefined'. See Defined provisional sums *and* Undefined provisional sums.

Public sector comparator (PSC) an assessment of the scheme which includes capital costs, operating costs and third-party revenues. The PSC is a benchmark against which value for money can be gauged. Clients use technical advisors to produce a reference project: sometimes called the public sector scheme (PSS).

Qualified tender where the contractor qualifies the tender in some defined way to reflect the contractor's concerns/requirements/suggestions (see *Variant bid).*

Residual risk (or retained risk) the risk that remains after all efforts have been made to mitigate or eliminate risks associated with the project. A risk assessment may identify a residual risk, but the risk is not completely controllable.

RIBA Plan of Work the definitive UK model for the building design and construction process. The Plan comprises eight work stages, each with clear boundaries, and details the tasks and outputs required at each stage.

RIBA Work Stage the stage into which the process of designing building projects and administering building contracts may be divided. Some variations of the RIBA Work Stages apply for design and build procurement.

Risk the likelihood of an uncertain event occurring, such as variation, an accident, additional costs, exceptionally inclement weather, price fluctuations. In a contract, items described as contractors' risk are those that the contractor takes on. They are additional technical, contractual, financial and managerial responsibilities which form part of the contractor's formal obligations.

Risk allowance the amount added to the base cost estimate for items that cannot be precisely predicted to arrive at an allowance that reflects the potential risk.

Risk register (or risk log) a list of risks (or opportunities), their value and probability of occurrence.

Scope opportunities to improve the financial, commercial or business aims of a construction organisation.

Selective tendering a method of selecting tenderers and obtaining tenders whereby a limited number of contractors are invited to tender. The tender list is made up of contractors who are considered suitable and able to carry out the work. This suitability is usually determined by pre-selection procedures.

Service level specification the specification given in the agreed project agreement setting out the standard to which the service must be provided. This is accompanied by an agreed performance monitoring regime.

Settlement the action taken by management to convert an estimate into a tender. Also commonly known as 'appraisal' or 'final review'.

Settlement meeting a timetable for the preparation of an estimate, all necessary supporting actions and for the subsequent conversion of the estimate into a tender.

Short-term programme the name given to the contractor's strategic work programme.

Site boundary the perimeter of the land and building that is under the control of the contractor during the construction and development period. The boundary forms the outer edge of the site area where construction operations take place without permission of neighbouring landowners.

Standing plant plant retained on site which is not working but for which a contractor is still liable.

Statutory undertaker organisations, such as water, gas, electricity and telecommunications companies, that are authorised by statute to construct and operate public utility undertakings.

Sub-contractor company or individual who is employed by the main contractor to undertake work. The work is part of the contract that has been awarded by the client to the main contractor under a subcontract; also known as specialist, works, trade, work package and labour only contractors.

Sub-contractor's preliminaries those preliminaries that relate specifically to building work to be carried out by a subcontractor. The associated costs should be included in the unit rates applied to sub-elements and individual components.

Sub-element group elements are divided into elements and further divided into sub-elements. These require separate cost targets to be established.

Temporary works resources needed for non-permanent work. Some temporary works such as formwork are measured in a bill of quantity, others such as hoardings are normally excluded from unit rate calculations because they are common to a number of activities and their durations are taken from the tender programme.

Tender a sum of money, time and other conditions required by a tenderer to complete the specified construction work. For design and build, the term tender includes design (Contractor's Proposals) and price (Contract Sum Analysis).

Tender Adjudication the process of converting the contractor's estimate into a tender bid. See Final review.

Tender inflation included as an allowance in the order of cost estimate or cost plan for changes in the basic prices of labour, plant, equipment and materials in the period between the estimate base date and the tender return date.

Tender documents documents provided for the information of tenderers, in order to establish a common basis for their offers.

Tender preparation programme resourced activity schedule outlining programme for preparation and submission of tender.

Tender settlement the conversion of a bid cost into a tender taking the commercial interests of the contractor into account. See Settlement.

Tender timetable a programme for the preparation of an estimate, which needs to show not only each item of work in the tender process, but also the person responsible and the associated times/dates.

Tender works programme a programme for the project resulting from the information available at the tender stage. It should not be relied upon for construction purposes but may be the basis for the construction master programme. Its purpose is to demonstrate that the contractor intends to comply with the data constraints listed in the tender documents and, where a contract period is not specified, it will indicate the completion date. Hence, the duration over which the time-related site costs must be included in the tender. It should be prepared as a critical path network in order to identify critical work activities and delivery dates of client-supplied information, goods and materials. It is not uncommon for the tender documents to require the tender works programme to be submitted with the tender.

Tendering a separate and subsequent commercial function based upon the estimate.

Time-related charge is for work, the cost of which is to be considered dependent on duration.

Total development cost the cost limit (including inflation i.e. the total of the works cost estimate, the project/design team fees estimate, other development/ project costs estimates, tender inflation and construction inflation) for the building project.

TUPE the Transfer of Undertakings (Protection of Employment) Regulations 1981.

Turnkey contracts a turnkey contract is a business arrangement whereby a project is delivered in a completed state. Therefore, rather than contracting with various parties to develop a project in stages, an client enters into a contract with one party (normally a developer or a contractor) to finish the entire project without any further input from the client. The developer or contractor is separate from the client, and the project is handed over only once it is fully operational. In effect, the developer or contractor is finishing the project and 'turning the key' over to the client. This type of arrangement can be used for construction projects ranging from single buildings to large-scale developments.

Unavailability the test for determining deductions from unitary payment by reference to standards for the provision of the facility for private finance initiative (PFI) projects.

Undefined provisional sum a sum typically used to make contingent provision for possible expenditure on elements of work which cannot be wholly foreseen at tender stage or cannot be quantified. A client's contingency sum is deemed to be an undefined provisional sum.

Unit rate(s) price applied per unit of works, goods or services (e.g. cost per m, cost per m^2 and cost per m^3). The term also includes costs/m^2 of GIFA and cost per functional unit (or functional unit cost) in a PPP/PFI arrangement.

Unitary payment the payment by the awarding authority to the project company for the provision of the facility under a PPP agreement.

Variant bid a bid which does not comply with the prescribed requirements of the awarding authority for a reference bid, but which a tenderer is proposing as offering better value for money.

Work breakdown structure (WBS) a task-oriented breakdown, which defines the work packages and tasks at a level above that defined in the networks and schedules. The WBS initiates the development of the Organizational Breakdown Structure (OBS) and the Cost Breakdown Structure (CBS). It also provides the foundation for determining earned value and activity networks.

Working Rule Agreement (WRA) national working rules for the building industry produced by the UK National Joint Council for the Building Industry.

Work package contractor a specialist contractor who undertakes particular identifiable aspects of work within the building project; for example, ground works, cladding, mechanical engineering services, electrical engineering services, lifts, soft landscape works or labour only. Depending on the contract strategy, works contractors can be employed directly by the employer or by the main contractor.

Works cost estimate the combined total estimated cost of the building works estimate, the main contractor's preliminaries and the main contractor's overheads and profit prepared using current prices at the time the estimate is prepared (or updated). The works cost estimate contains no allowance for project/design team fees, other development/project costs, risk allowances, tender inflation and construction inflation

Works package contractor's preliminaries preliminaries that relate specifically to the work that is to be carried out by a works package contractor.

Code of estimating and tendering practice – principles and procedures

March 2018

Contents

Foreword

The CIOB's New Code of Estimating Practice (CoEP) builds upon previous versions of the CIOB Code of Estimating Practice, and provides members and other professionals with summary guidance for the estimating process and accepted good practice.

The CoEP reflects the changing practices of estimating, pricing, bidding and tendering for work in the construction sector. It sets out the underlying principles that can assist estimators in pricing projects, and professionals engaged in pricing and bid preparation to establish and maintain a set of procedures for estimating, tailored to the needs of their particular company. The principles are equally applicable to public and private sector building and civil engineering works, and for use by micro, small, medium and large constructors for small to large projects. It recognises the importance of all the stakeholders in the construction process.

Estimates/bids require the highest principles of ethical conduct to ensure the protection of the public, clients, employers and others in the industry and related professions. Good practice in a competitive environment is essential to maintain the professionalism of estimating, something the Institute takes very seriously.

Chris Blythe OBE
Chief Executive, CIOB
March 2018

1 Introduction and scope

1.1 This Code of Estimating and Tendering Practice embodies the principles and procedures outlined in the New Code of Estimating Practice. It sets out a best practice guide.

1.2 The New Code of Estimating Practice provides guidance to construction organisations, consultants and clients/owners/employers/project sponsors for all types and sizes of projects:

(a) Estimators that undertake the pricing working for constructors,[1] specialty contractors, suppliers and specialists in the supply chain.

(b) Consultants who provide forecasts and estimates for proposed construction work.

(c) Clients[2] who procure projects, need an understanding of how the estimating process produces the final price and seek confidence that any estimate has been professionally prepared to the highest possible standards.

(d) Educators and trainers who need to understand the detailed issues in estimating and the broad context in which the discipline is applied.

1.3 The term 'tender' and 'bid' are often used interchangeably. However, 'bid' is increasingly used by the offerer (the supply side), and the term 'tender' used on the procurement side (the buyer).
The bid is the process of preparing the price for the work; the tender is the formal offer.

1.4 The tenderer is the person or organisation bidding for the work; this includes the terms builder, service provider, constructor, lead contractor, principal contractor, general contractor and supplier. The client must treat a tenderer equally, openly, fairly and honestly. The principle of tendering is to ensure that true competition is achieved, comprising a simple price assessment or more complex evaluation embodying quality, safety, environmental, ethical, technical and other factors.

1.5 The tendering process must be competitive, equitable, fair, transparent and cost-effective.

1.6 Bidding/estimating is not a costless activity. The cost of bidding can be very substantial, which must be borne in mind by procurers and every stakeholder in the construction design and delivery process. Cost-effective procurement must reflect the cost, time and resources involved. Ultimately, the client pays for abortive estimating with the costs recovered in the project overheads. The larger the tender list, the greater will be the cost of abortive tendering, reflected in higher prices. Due regard should be taken of the amount of work demanded of tenderers in order to formulate their bids, which involves extensive quantification, specification, specialisation and/or calculation, plus the cost and time of the specialty contractors and suppliers in preparing estimates and programmes of work.

[1] The term constructor has been used to represent the contractor/general contractor/principal contractor/lead contractor/works contractor.

[2] The term client has been used throughout to refer to the project client/owner/employer/project sponsor.

1.7 Good/accurate/reliable estimating is fundamental to the success of any construction enterprise to ensure sensible prices and reasonable profit margins. The accepted tender establishes the price for the project; the constructor's production team will live with the consequences of an unreliable/inaccurate bid throughout the duration of the site production phase. Consistent and predictable results are achieved more effectively and efficiently when activities are understood and managed as interrelated processes that function as a coherent system within the estimating process. Clients want sensible and accurate prices that they can rely on; they do not want surprises.

1.8 The constructor shall comply with all legal obligations and establish a code of behaviour for estimating, bidding and tendering to regulate the actions of its employees. Such a code shall at least require that these persons:

(a) Discharge their duties and obligations on time and with integrity to all stakeholders.

(b) Behave equitably, honestly and transparently.

(c) Avoid conflicts of interest and, where a conflict of interest is known, declare and address that conflict.

(d) Do not maliciously or recklessly injure, or attempt to injure, the reputation of another party.

2 Procurement

2.1 Procurement is the overall act of obtaining goods and services from external sources. It includes deciding the strategy to solicit tender offers, how those goods are to be acquired by reviewing the client's requirements (i.e. time, quality and cost) and how the tenders will be evaluated.

2.2 All procurement must conform to the three pillars of Integrity, Transparency and Accountability.

2.3 The five main methods of procurement are:

(a) Traditional/conventional, where design is separated from construction.

(b) Design and build, Turnkey.

(c) Management procurement, such as construction management and management contracting where design and production can proceed in parallel.

(d) Integrated, sometimes known as collaborative procurement, partnering and alliancing, where the focus is upon collaboration and working together.

(e) Concession agreements, such as build/operate/transfer (BOT), public–private partnerships and private finance initiatives, where a team is formed for project delivery over a concession period awarded by the sponsor.

2.4 BS ISO 10845-1:2010 (Construction procurement policies, strategies and procedures. Code of practice) defines the principles of procurement, and BS ISO 10845-2:2011 (formatting and compilation of procurement documentation). Consideration should be given to:

(a) Methods of selection of the preferred bidder, whether by lowest price, most economic advantaged tender (MEAT) or other selection criteria.

(b) Issues of bribery and anti-competitive practice.

(c) Dispute resolution.

(d) Methods of identification and management of risk.

(e) Issues of payment and financial management.

(f) Corporate social responsibility.

(g) Health and safety.

(h) Environmental sustainability.

(i) Intellectual property.

(j) Conflict of interest.

(k) Attitude to risk allocation.

2.5 In addition, consideration must be given to:

(a) Public procurement rules.

(b) Taxation legislation and Value-Added Tax.

(c) Planning and statutory requirements that impact the tender.

(d) Levies, taxes and duties payable.

2.6 In summary, procurement is about:

(a) Establishing what is to be procured.

(b) Deciding on the procurement strategy.

(c) Soliciting tender offers.

(d) Evaluating tender offers based on the defined selection criteria.

(e) Awarding contract.

2.7 The tender documents should clearly define the roles and responsibilities of all the stakeholders.

2.8 The criteria for tender evaluation and the methods for applying such criteria in evaluation must be clearly stated in the tender documents to ensure that all the parties understand the basis of the selection criteria.

2.9 The scope of projects, which is provided by the client or their consultant, must be sufficiently described to create certainty about work scope at the tender stage. Bidders should avoid the practice of hoping to raise claims during production because of incomplete information.

2.10 If the time for project completion stipulated in the tender documents is unrealistic and based on policy expectations of the client, these expectations should be contested by bidders. Whilst tenderers want to avoid disqualification from the tendering process, it is important that dialogue takes place.

3 e-Procurement

3.1 e-Procurement provides for every stage of the process an application that allows the electronic handling of the tender, starting from the publication of the contract notice up to the awarding of the contract.

3.2 Tender documents can be submitted electronically, provided the client allows such practice. This can be done via e-tendering, allowing companies to submit their tender or request to participate electronically on the project.

3.3 Once all the documents have been uploaded, the submission report must be signed. This report features all relevant details about the documents uploaded.

3.4 Where it is indicated in the Invitation to Tender that a tender can, or must, be submitted electronically, then those tenders shall be:

(a) addressed to the e-mail address as notified in the Invitation to Tender

(b) in the format specified in the Invitation to Tender

(c) stored in a secure mailbox, which requires a code or other appropriate security measure to open it

(d) retained unopened until the date and time specified for its opening.

3.5 Where tenders are submitted electronically or by other digital media, appropriate systems should be in place to ensure that receipt can be clearly recorded to ensure that tender timescales are complied with.

4 Pre-qualification

4.1 Pre-qualification is a necessity for public sector and non-governmental organisations (including central government authorities, local authorities, universities, NHS Trusts (UK) and some utility companies) and with some large private sector clients with an ongoing programme of work. It is used for the establishment by the client of a list of constructors or specialty constructors with the necessary skills, experience, resources, previous tender performance and desire to carry out the works, bearing in mind the character, size, location and timing of the project.

4.2 Pre-qualification will occur before any formal invitations to tender are issued.

4.3 The client/consultant should allow a realistic programme covering the whole of the pre-qualification and tendering period and adequate time given for each stage.

4.4 BSI PAS 91:2013 (Construction prequalification questionnaires) is a publicly available specification that sets out the content, format and use of questions that are widely applicable to pre-qualification for construction tendering. It states: 'To be eligible for pre-qualification it is necessary that suppliers demonstrate that they possess or have access to the governance, qualifications and references, expertise, competence, health and safety/environmental/financial and other essential capabilities necessary for them to undertake work and deliver services for potential buyers.... The use of this set of common criteria by those who provide pre-qualification services will help to streamline tendering processes.'

4.5 Pre-qualification of a joint venture formed to bid for a project is an important requirement. A joint venture can be described as a business enterprise where two or more participants come together to share their expertise in order to win a specific contract for a set period of time. They collaborate to bid for a project. The advantages of the joint venture are that the skills sets of the participants involved in the joint venture are combined, the risk exposure is shared and the participants share the profits as well as the costs in the joint venture.

A joint venture is a separate entity and must have its own set of rules by which it is managed, called a joint venture agreement, which must include some minimum rights and duties:

(a) The way the profit or loss will be shared by the participants.

(b) The duties of each participant towards the joint venture.

(c) The reason the joint venture has been established.

(d) The start and end date of the joint venture.

(e) The persons who will be the representatives of the joint venture.

4.6 The tender documents must stipulate the documentary requirements for any joint venture tender and the conditions for acceptance of the tender offer.

5 Types of tender

5.1 The types of tenders are varied; the most common are single-stage (where the design has been sufficiently developed to enable pricing of the work), two-stage and competitive dialogue tendering procedures.

(a) Open – the project is advertised publicly in the press and on the Internet; any organisation can apply for the documents and to bid, provided they meet the selection requirements. The advantage is the wide range of companies likely to be interested in bidding. The disadvantage is that open tendering is indiscriminate and likely to attract bidders who are not suitably qualified for project.

(b) Selected/qualified – a selected number of suitably-qualified contractors are invited to submit a bid based on pre-qualification. The number of bidders should not be excessive, normally no more than five for a major project.

(c) Negotiated – a single company produces a price for the work as the sole bidder. Trust is an important part of negotiation with the contractor; frequently, an owner has developed an ongoing relationship with a company. Negotiation is involved in a two-stage tender process where the successful contractor from the first stage negotiates the second stage of the bidding process.

(d) Two stage – two-stage tendering is a hybrid approach, usually for design and build, that seeks to exploit the advantages of both negotiation and competition. It accelerates the process by permitting the overlap of design and procurement. The appointment of a contractor is in two stages: stage 1 is competitive and based on costs for preliminaries, overheads and profit; the stage 2 appointment is made after a satisfactory negotiation of the final price.

(e) Serial – used by the client when they have a number of similar projects. The pricing of the first project can lead to a number of sequential projects. This type of tendering can lead to economies of scale and efficiencies. The contractor has familiarity with the client requirements and the design issues.

(f) Framework – involves invitations to tender to contractors that operate a framework agreement with the client where there is agreement between a client and one or more contractors, the purpose of which is to establish the terms governing contracts to be awarded during a given period. The framework normally is for a fixed period. A number of contractors can be selected for the framework.

6 The decision to tender

6.1 Factors considered in the decision to submit a bid:

(a) Cost of preparing the bid ranked against the likelihood of success.

(b) The competition for the project and the likely number of bidders.

(c) Time given in the tender to prepare the bid.

(d) Workload and capacity of the estimating department.

(e) Financial situation with cash flow and capital requirements to undertake the project.

(f) Operational capacity with experience and competencies required.

(g) Strategic direction of the business and the fit of the potential project.

(h) Any conflict of interest.

(i) Realistic duration times for the work if specified by the client.

(j) Experience of working with the client/design consultants and the working relationship.

7 Tender documents

7.1 Inspection of the tender documents must seek to achieve, as a minimum, the following objectives:

(a) The documents and information received are those for the project under review; they are adequate for assessing costs and the risk.

(b) Sufficient time is available for production and delivery of the tender. The bidder should notify the client at the earliest opportunity that there are concerns on the time allocated for the tender process where insufficient time has been allowed.

(c) The drawn information is sufficiently well developed to use for reliable pricing with no significant areas of uncertainty or unreasonable assumptions that must be made because of lack of information.

7.2 The tender documents on a traditional/conventional contract of design–bid–build usually comprise:

(a) Notice to tenderer.

(b) Form of tender.

(c) General conditions of the contract/terms of engagement, terms that collectively describe the rights and obligations of contracting parties and the agreed procedures for the administration of the contract or document containing conditions of contract.

(d) Specification of the works with details of any performance requirements.

(e) Drawings that form the basis for the tender.

(f) Bills of Quantities produced by the independent cost consultant appointed by the client or details of measurement and price information that must be submitted with the tender if no bill of quantities has been provided.

(g) The evaluation criteria for the selection of the preferred tender.

7.3 The quality of the tender documentation must be assessed objectively, thus avoiding lack of information and unreliable information, which could lead to disputes and contractual claims. Enough information, such as clearly defined elements of work, will reduce the level of intuition/guesswork/assumptions/inaccurate allowances for risk and uncertainty. Where there is a lack of clarity on the drawn design information, the bidder should submit a request for more information.

7.4 The tender document must state clearly either the construction duration from site possession to practical completion and handover or request the bidder specify a construction duration.

7.5 A tenderer requiring any clarification of the tender documents shall contact the Client/or Client's consultant in writing or raise the enquiry during any pre-tender meeting, if such a meeting takes place. The Client will respond to any request for clarification, provided that such request is received prior to the deadline for submission of Tenders, within a reasonable period. The Client's response shall be in writing with copies to all tenderers.

7.6 The tender offer is a written offer for the provision of goods or to carry out a service for engineering and construction works under given conditions, usually at a stated price, which is capable of acceptance and conversion into a binding contract.

7.7 Incomplete, inadequate and poorly detailed design information at the tender stage costs money for the contractor and ultimately the client – the risk allowances that must be added to the bid may not be correct. It is a false economy to start a project with poor information.

7.8 Over-long tender lists result in unnecessary abortive costs for clients and tendering organisations; tenderers might put less effort into their tender submission if they are one of many. A short tender list ensures that tenders are received from the most suitable organisations.

8 Bid preparation and pricing

8.1 Pricing a project is about:

(a) The allocation of physical, financial and human resources to the items/activities specified on the drawings, the specification and the bill of quantities.

(b) Productivity factors related to output or production.

(c) Making an allowance for a fair and acceptable level of compensation and profit.

(d) Making due allowance in the estimating for uncertainty and risk associated with the work.

(e) Making allowances for future price inflation of labour, materials, plant and equipment.

(f) Managing and allocating risk to the party best equipped to handle it.

(g) Taking account of the market conditions and the buying environment in the construction sector.

(h) Recognising the work packages and specialty contractors that will undertake the work items and the time and information needed to estimate the work package.

(i) Incorporating the prices of the suppliers of materials, plant and equipment, with allowances for their special requirements.

(j) Making allowance for wastage of materials and the cost of removal of waste from site, including the disposal costs.

(k) Ensuring that good commercial decisions reflect the risk involved.

(l) Incorporating the cost implications of items in the Working Rule Agreement for the Construction Industry, with due allowance on construction prices.

(m) Calculating the cost implications for compliance with the conditions and terms of contract.

(n) Undertaking the work within the stipulated contract duration or estimating a contract duration.

(o) Consideration of the method and programme of work.

(p) Consideration of what temporary works are required.

(q) Incorporating the cost of temporary site utilities (e.g. water, power and communications).

(r) Estimating the preliminaries items using resources, project duration and method of working.

(s) Making allowances for the cost implications of occupational health and safety, with personal protective equipment, training and safe systems of working.

(t) Including proper allowances for site-based and head office overheads.

8.2 Each estimate used in the bid should be cross-checked to ensure that it is technically and mechanically free from mistakes, oversight or errors. With an estimated price, the estimator is stating that the estimate has been prepared to the best of their ability using their education, expertise, judgement, knowledge, skills and recognised standards, and based on the available information. When possible and the size of project warrants it, peer review of the tender submission should be undertaken by another internal 'team' prior to submission.

8.3 Poor estimating can lead to huge financial and reputational losses on a project. On large projects, estimating can involve thousands of individual estimates for work packages that take account of variables including the size, scope, complexity and locality of a specific project. The more detailed the estimate, the better the accuracy/reliability. Competent, empowered and engaged people at all levels throughout the process of estimating prices are essential to enhance the constructor's capability to create and deliver value.

8.4 Bidding should adopt and observe the key values of fairness, clarity, simplicity and accountability as well as reinforce the idea that the apportionment of risk is fundamental to the success of a project.

9 The tender

9.1 Tendering involves more than simply obtaining a price. Tendering is:

(a) The bidding process to obtain a price, duration and details of the delivery process, including policies on health and safety, quality, sustainability and how the constructor is going to deliver the project to meet the clients expectations.

(b) Selection and appointment of the successful contractor following tender submission and review.

9.2 An invitation to tender is usually an invitation to treat.

9.3 A tender is in the nature of an offer, which in the legal sense is capable of acceptance. A tender bid is submitted to the client in the prescribed form; it is an offer to undertake the project. The tender offer can be withdrawn at any time up to the acceptance.

9.4 The robustness/reliability/accuracy of the tender depends on the level of design and specification detail available for the production of the bid. Poor-quality bid documents lead to unreliable bids, which will be reflected in the risk allowances and contingencies that must be included to cope with uncertainty and risks.

10 Bid timetable

10.1 The more complicated the works, the longer the tender process will take.

10.2 The bid timetable must recognise the complexity of the task and the likely number of organisations that will be required to submit offers to the constructor.

For single-stage tenders on a straightforward project, a minimum of 28 days should be allowed for the tender period, recognising that the contractor will be bidding on more than one project. The bidding period should be increased for projects that are more complex.
For public sector projects, a minimum of 40 days should be allowed.

10.3 The bid timetable highlights the key dates in the production of the tender.

(a) Latest date for dispatch of enquiries for materials, plant and specialty contracted items.

(b) Latest date for the receipt of quotations.

(c) Key dates for bills of quantities production, drawings and specification for Design and Build projects.

(d) Visit to the site and the local area with consideration of site access and any restrictions.

(e) Preparation of the method statement.

(f) Preparation of the tender construction programme.

(g) Completion of pricing the measured unit rates.

(h) Intermediate co-ordination meetings for the bid team.

(i) Identification of key personnel who will be involved in site management and production management.

(j) Review meetings and requests for further information form the design team.

(k) Tender adjudication meeting prior to submission

(l) Submission of the tender.

10.4 Personnel associated with the tender must confirm that they are able to provide the necessary data and information in the format required, in accordance with the bid timetable.

10.5 Adequate time is required for the assimilation of project information, obtaining quotations from specialty/trade constructors and suppliers, and for completing the pricing of the bid.

11 Enquiry documents to specialty constructors/suppliers

11.1 A list of items for each enquiry package is made by collating items electronically from the tender documents. This list can be used while assembling the package, as a record of what enquiry documentation was sent. It should be reproduced in the enquiry document so that the recipient can check that all the relevant parts have been received.

11.2 Work/trade/contract/supply packages are awarded to specialty contractors/suppliers. The specialty contractors/suppliers are treated as stakeholders in a fair and equitable manner. They should be notified of the position of their bid and kept informed of progress. The requirement is to:

(a) Minimise the number of interfaces between packages.

(b) Place and define required interfaces between contract packages so that they are relatively easy to control and manage.

(c) Analyse interfaces rigorously to ensure that there are no gaps when pricing the works.

(d) Enable use of different contract conditions to suit the different elements of work.

(e) Enable the selection of 'construction only' for some contracts and design and build for others.

(f) Allow suppliers to be selected to match the skills and capabilities required.

(g) Enable risks to be allocated differently between constructor and supplier for different work packages, placing risks with those organisations most able to manage them and thus control the costs.

(h) Sub-divide work so that more suppliers have the ability to price for it, spreading the risk of delivery and/or to enable smaller suppliers with lower cost structures to undertake elements of the work cost-effectively.

11.3 Bid shopping, as known in the construction industry, is not good estimating practice. Bid shopping occurs when after the award of the contract, a constructor contacts several specialty contractors of the same discipline in an effort to reduce previously quoted prices. This practice is unethical, unfair and not in the best interests of any of the parties. Whilst seeking the lowest price may be accepted practice, the industry should move towards the most economically advantageous price and best value for everyone concerned.

12 Risks

12.1 Risk allowances need to be included in the bid. Once identified, they are included with an allowance in the bid prices. They may be subject to adjustment by the management for commercial/competitive reasons at the tender adjudication stage.

12.2 Risks are those to be borne by the client and the constructor. In each category, they are classified as known risks, known unknowns (uncertainties) and unknown unknowns (e.g. force majeure). Risk allowances may be lump sums or an allowance based on a percentage of the price.

13 Qualifying the tender

13.1 Qualifications, or tender notes, can be used to set aside items that the estimator believes should be excluded and to limit risks, for instance by giving provisional sums for unclear items rather than adding a risk allowance to the tender (should the tender document allow). Such strategies, if clearly explained, are legitimate and often necessary in competitive tendering. The risk and implications of a non-compliant tender should be considered at the tender settlement stage by the constructor's senior management.

13.2 The Best and Final Offer (BAFO) is the term used to indicate that no further negotiation on the amount or terms is possible. A BAFO is issued often in response to the request of a client to those contractors or suppliers whose bids are within a close range of one another.

14 Bid submission and evaluation

14.1 The selection stage of the process should be objective, fair, accountable and transparent. There should be no deviation from the stated evaluation criteria in the tender documents.

14.2 If requested, the contractor will supply details of:

(a) Procedures for planning, programming, scheduling and management.

(b) Key personnel for site management and project management.

(c) Mobilisation and construction schedules from contract award, pre-production and site commencement.

(d) Procedures for safety and health, and well-being, of the workforce.

(e) Programme for completing the project, the tender programme, including milestones for achieving objectives.

(f) Site organisation structure.

(g) Method statement outlining the sequence and method of working with appropriate plant.

(h) Identified risks and proposals for their management.

(i) Communication arrangements for the job site.

(j) Quality plan to ensure compliance with the required quality standards.

(k) Policy on sustainability.

14.3 Tender submissions, except those which the client permits to be submitted by telefax or by electronic means, shall only be accepted for evaluation when they are submitted in sealed envelopes, annotated with the required particulars and placed in the nominated tender box or delivered to the specified place for receipt of tender submissions (BS ISO 10845-1:2010).

14.4 The invitation to tender document should include provisions stating that tenders can be rejected if they are not compliant with the requirements, including compliance with submission dates, times and format.

14.5 Justification to reject a tender must be based on the existence of one or more major deficiencies or deviations, which cannot be permitted to be rectified or accepted. A material deviation is one which:

(a) has an effect on the validity of the bid; or

(b) has been specified in the bidding documents as grounds for rejection of the bid or

(c) is a deviation from the contractual terms or the technical specifications in the tender documents.

14.6 Late tenders should normally be rejected unless:

(a) late delivery is a result of actions outside the control of the tenderer,

(b) other exceptional circumstances exist which the client, in exercising reasonable discretion, deems sufficient to allow acceptance.

(c) Where a decision is made to accept a late tender, the reasons why the tender has been accepted must be clearly stated.

14.7 It is important that bidders, clients and consultants preserve commercial confidence when handling tendering information. The bidder's prices are confidential and should not be shared with any party without prior agreement in writing.

15 Errors and Omissions in the tender

15.1 Where a bill of quantities or a schedule of prices is used:

(a) The price submitted (i.e. the offer), which is made known to interested parties at the opening of tenders, shall be used as the basis for establishing the competitive position of tenderers.

(b) The most competitive tender shall be checked for arithmetical errors, where relevant, and where such errors are found, the tenderer shall be notified of the errors and invited to either confirm the tender offer as tendered or accept the corrected total of prices.

(c) Where the correction of the errors results in a change in the competitive position of tenderers, the process in (b) above shall be repeated.

15.2 Where an error is discovered in the pricing of the works, Joint Contracts Tribunal Ltd (JCT Practice Note: Tendering 2017) recommends that two alternatives should be stated in the tender documents:

(a) Alternative 1 – The tenderer should be given details of the errors in the tender and given the opportunity to stand by or withdraw the tender. If the tender is withdrawn, then the price or next best value tender of the next lowest tenderer should be considered. A request to withdraw is deemed an acknowledgement by the bidder or proposer that it is unwilling to undertake the project pursuant to the bid or proposal.

(b) Alternative 2 – the tenderer should be given the opportunity to confirm the tender offer or correct genuine errors.

16 Contract award

16.1 The tender documents must clearly state the evaluation process and the extent to which they are mandatory, desirable or optional parts of the requirements.

16.2 A contract should be awarded based on either:

(a) Lowest price: the lowest financial offer. No other element of the tender is taken into account.

(b) Obtains the highest number of tender-evaluation points, sometimes called 'The most economically advantageous tender' (MEAT), or best value criteria: Factors other than or in addition to price, like quality, technical merit and running costs can be taken into account. In MEAT, the contract award criteria, for example price, quality of services, risk to contracting authority and any sub-criteria must be set out in the tender documents, and the weighting of each criterion, and sub-criterion (if weighted), must also normally be given, either as an exact number or as a meaningful range (e.g. 'price: 30–40%').

(c) Value for Money is a core procurement principle. This means a shift from the lowest evaluated compliant tender to tenders that provide the best overall value for money, taking into account quality, cost and other factors as specified.

16.3 The most economically advantageous tender identifies the bidder that offers best value for money and compares bids on commercial, technical and financial terms, i.e. not just on price element, and to allow for optional requirements to be priced and evaluated on a lifetime cost basis, that is allows each criteria to be shown as either 'mandatory' or 'optional'.

16.4 When the evaluation of the tender has been completed, the client/consultant shall check if the evaluated cost is reasonable, whether it may be identified as unbalanced, front loaded or abnormally low.

16.5 Under English Law, a tender may be withdrawn any time before it has been accepted. Once a contract has been awarded, all the unsuccessful tenderers should be informed to release them from any obligations on their resources.

16.6 The tender will remain open for acceptance, normally for 60 days, following the submission date. The client does not have to accept the lowest or any tender submitted.

16.7 After the successful tenderer has received the client's notice of acceptance, all other tenderers are notified that their tender offers have not been accepted by post, telefax or another electronic method, or by publication of the name of the successful tenderer on a website, or in an accessible publication.

16.8 In the event that a tenderer becomes ineligible for the award of a tender, the negotiations result in the competitive position of a tenderer changing, or the negotiations fail to reach mutual agreement, the next highest scoring or highest ranked tenderer shall be declared the preferred tenderer. The procedure shall be repeated until such time that a contract is concluded (BS ISO 10845-1:2010).

17 Ethical standards in tendering

17.1 Tenderers, suppliers, contractors and specialty contractors should observe the highest standard of ethical behaviour during the bid and tender process. Prohibited practices are:

(a) 'Corrupt practice' means the offering, giving, receiving or soliciting, directly or indirectly, of anything of value to influence improperly the actions of another party.

(b) 'fraudulent practice' means any act or omission, including a misrepresentation, that knowingly or recklessly misleads, or attempts to mislead, a party to obtain a financial or other benefit or to avoid an obligation.

(c) 'Coercive practice' means impairing or harming, or threatening to impair or harm, directly or indirectly, any party or the property of the party to influence improperly the actions of a party.

(d) 'Collusive practice' means an arrangement between two or more parties designed to achieve an improper purpose, including influencing improperly the actions of another party.

17.2 Anti-competitive behaviour is prohibited under Chapters I and II of the Competition Act 1998 (the Act) and may be prohibited under Articles 101 and 102 of the Treaty on the Functioning of the European Union (TFEU). These laws prohibit anti-competitive agreements between businesses and the abuse of a dominant position in a market.

Section One
Principles – the theory and background

The Principles section is made up of 16 chapters which can be accessed by selecting a square in the graphic below. The order of the chapters is not significant; they are designed to be stand-alone, providing a reference document for the reader.

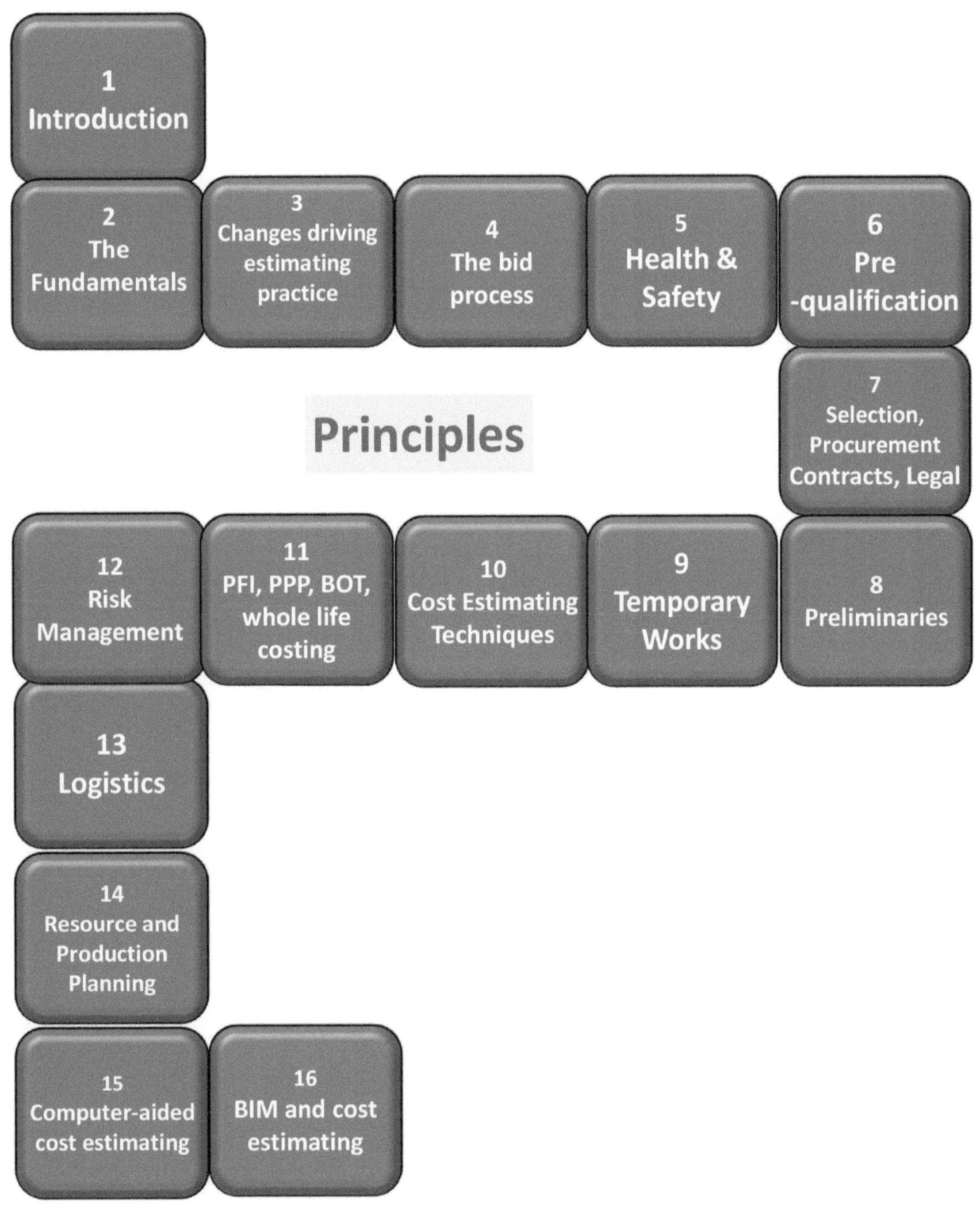

1 Introduction

Estimating *is the systematic analytical calculation of projected labour, material, plant, equipment and overhead costs for a project. The aim is to produce the most accurate, reliable and best estimates, inclusive of contingencies reflecting the project risks, using the most reliable and complete project and pricing information available at the time the estimate is prepared.*

This New Code of Estimating Practice (CoEP) provides users with recommendations for accepted good practice. It reflects the changing practices of estimating, pricing and bidding for work in the construction sector. The CoEP sets out procedures and practice that can assist estimators in pricing the work and other bid preparation professionals to establish and maintain a set of procedures for estimating, tailored to the needs of their particular company.

Some of the procedures set out may be unnecessary or too complex for some organisations because of the nature of the work being undertaken. However, the underlying principles are generally applicable and should be borne in mind. The principles are equally applicable for public and private sector building and civil engineering works and for use by micro, small, medium and large contractors on small, medium and large projects.

Estimating is involved from the strategic definition stage of a project, through the design, tendering and production stage, up to completion and use. It ensures construction organisations can survive in the highly competitive market place by securing work at the right price, whilst the risks have been considered and taken into account.

English Dictionary definition of estimate: roughly calculate or judge the value/number/quantity/extent of, approximate, make an estimate of, guess, evaluate, judge, gauge, reckon, rate, appraise, form an opinion of, form an impression of, get the measure of, determine, weigh up.

Every organisation involved in construction from contractors to suppliers and manufacturers must have good, robust, reliable, accurate, estimating and feedback systems in order to survive.

1.1 An imprecise science

Estimating cannot be a precise science; it deals with uncertainty and many unknowns. Assumptions have to be made about the weather, ground conditions, productivity factors, inflation and the mitigation of risk.

People are the most important element; computer systems and software have helped, but lack of information will inevitably cause difficulties for the estimator. Estimators/bid managers* have to be innovative and creative in the way they address tender

* Estimators have the specialization of formulating prices for bids. Some companies use the bid manager to manage the bid team.

New Code of Estimating Practice, First Edition. The Chartered Institute of Building.

The terms cost, price, estimating and cost estimating are used extensively in the New CoEP. Cost is the expense that an organisation incurs in bringing a product or service to market. Price is the amount a customer pays for that product or service. The difference between the price that is paid and the cost that is incurred is the profit. Cost is often split into two elements: the variable cost (costs that vary in proportion to the turnover) and fixed costs (costs that are required, but that do not vary as output varies). An estimate is the judgement of the resources required based upon the best information available.

responses. Every activity in the tendering process has a time, cost, quality and process implication for job site production and project delivery.

Pricing a project is about:

- The proper allocation of physical, financial and human resources to the items/activities
- Understanding and taking uncertainty and risk into account
- Ensuring good commercial decisions reflect the risk involved
- Managing and allocating risk to the party best equipped to handle it
- Ensuring compliance with the conditions and terms of contract and the cost implications
- Undertaking the work within the duration (if stipulated)
- Consideration of the method and programme of work
- Consideration of what temporary works are required
- Consideration of the works packages to be allocated to a specialist contractor
- Including proper allowances for site based and head office overheads
- Making an allowance for a fair and acceptable level of compensation and profit
- Making allowances for price inflation of labour, materials, plant and equipment
- Taking account of the market conditions and the buying environment in the construction sector.

The information you have is not the information you want, the information you want is not the information you need, the information you need is not the information you can obtain, the information you can obtain costs more than you want to pay.
(Bernstein 1998)

Good estimating is fundamental to the success of any construction enterprise; ensuring work is undertaken with sensible prices and reasonable profit margins. The accepted tender establishes the price for the project; the contractor's production team will live with the consequences of an unreliable/inaccurate bid throughout the duration of the site production phase.

Bidding should adopt and observe the key values of fairness, clarity, simplicity and accountability as well as reinforce the idea that the apportionment of risk is fundamental to the success of a project.

Trust is important when bidding with reliance upon the strength, integrity, ability and surety of every organisation involved in project delivery, confident that they are reliable and trustworthy.

Poor estimating can lead to huge financial and reputational losses on a project, even leading to job losses. On large projects, estimating can involve thousands of individual estimates for work packages that take account of variables including the size, scope, complexity and locality of a specific project.

1.2 Data, information, and knowledge in estimating

The quality and completeness of the design and specification information at the tender stage is crucial to ensure the quality and accuracy of bidding. Poor-quality information leads to poor-quality bidding; lack of data and information means the contractor must add large risk allowances to cope with uncertainty and the unknowns.

For design and build projects, the scope of work and the client requirements should be clearly defined. Key dates for delivery, quality and performance requirements should be clear and unambiguous. Ambiguity in the scope or specification of the work will lead to problems for both the employer and the contractor.

A simple definition of knowledge is the information, understanding, comprehension and skill, obtained through education, training, familiarity and experience. Knowledge is what we know; it also contains beliefs and prejudices.

1.3 Experience, instinct, gut feel, intuition and bias

A collection of data is not information. A collection of information is not knowledge. A collection of knowledge is not wisdom. A collection of wisdom is not truth.

Instinct, gut feel and intuition relate to behavioural decision-making – how decisions are made by estimators and members of the construction team. People make decisions that are impacted by heuristics (the application of experience-derived knowledge to a problem) and bias.

The 'intuitive' decision is based solely on a personal feeling of rightness or inevitability.

Bias indicates that information has not been processed objectively and instead has been interpreted in a subjective manner according to the particular 'prejudices' and 'influences' of the individual or decision-maker. Each of the many other people/groups that the estimator deals with at the bid stage have their own biases and interpretation of the terms used in the assessment of probability of their estimates being correct or otherwise – see Fig. 1.1.

Intuition involves the use of a sound, rational and relevant knowledge base in situations that, through experience, are so familiar that the person has learned how to recognise and act on appropriate patterns. All estimating involves knowledge, experience, instinct, gut feel and intuition, combined with tools and techniques, applied to reliable information about what is required.

Errors in estimating, prediction and forecasting are unavoidable; the world is not predictable. Any forecast about the future is based upon the best information available. Even though evidence shows that judgement is less accurate than statistical data, people continue to rely on their judgement.

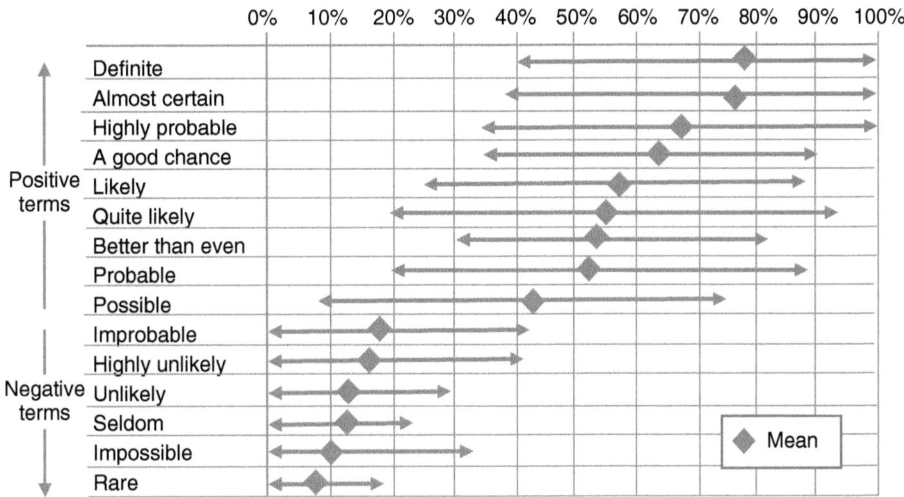

Figure 1.1 The interpretation of probability-related terms. *Source:* Hillson and Hulett (2004).

1.4 Optimism bias

People have an innate bias towards optimism, called optimism bias. Estimators must be aware of the trend towards optimism bias and the desire to win the project in competition. Having estimates adjusted by a group of people can be dangerous as most people don't feel comfortable going against group decisions; this is called group bias.

Figure 1.1 shows the distribution of meanings of probability terms across a range of respondents. The variability is significant in some instances, such as 'possible', 'probable' and 'quite likely'. The distribution is more marked in the positive, rather than the negative terms. Estimating involves many assumptions and judgements about likelihoods.

2 The Fundamentals

2.1 Tender and bid

The term 'tender' and 'bid' are often used interchangeably. However, 'bid' is increasingly used by the offerer (the supply side), and the term 'tender' used on the procurement side (the buyer).

An invitation to tender is usually an invitation to treat. A tender is in the nature of an offer, which in the legal sense is capable of acceptance. A tender bid is submitted to the client in the prescribed form; it is an offer to undertake the project. The tender offer can be withdrawn at any time up to the acceptance.

The tenderer is the person or organisation bidding for the work; this includes the terms builder, service provider, contractor and supplier.

A tenderer is required to be treated equally, openly, fairly and honestly by the employer.

The principle of tendering is to ensure that true competition is achieved, comprising a simple price assessment or more complex evaluation embodying quality, safety, environmental, ethical, technical and other factors.

2.2 Pricing

Once a tender is accepted and a formal contract executed, the successful bidder will become the contractor for the purposes of the contract.

'...a successful project manager/site manager is only as good as the information they are given to work with.'

Pricing a project at the tender stage involves many people with:

- Systems and processes
- Analytical tools
- Data (measurement, specification, drawings, pictures and verbal)
- Information
- Technical skills
- Assumptions
- Forecasts
- Knowledge
- Experience
- Professional judgement
- Intuition.

There must be reliable input, a system and output.

New Code of Estimating Practice, First Edition. The Chartered Institute of Building.

Reliable input means completed drawings, specification that clearly defines the scope and quality of the work, measured work and conditions of contract, with the duration for the activity or the project stipulated, or left to the discretion of the contractor.

Estimates and bids must be accurate, reliable and correct; however, there will be margins of error when forecasts, assumptions and allowances for risk are included in the estimating.

The term 'estimate' should be used carefully, to ensure that the recipient is clear whether the estimate constitutes a formal offer, or the estimate is an indication of likely cost. There is no rule of law that the use of the word 'estimate' on a document prevents it amounting to an offer. Where such a document carries standard terms and conditions, that will generally be a strong indication that the document is an offer capable of giving rise to a contract if accepted expressly or by conduct.

Pricing construction activities involves understanding the measured item with:

- The description of the activity
- Its context in a project (is it a single activity or part of multiple activities?)
- The location of the activity on the project
- The resources and specialist skills needed to undertake the activity
- The productivity rates
- Production methods to be used
- The time required for the activity
- The relationship and interdependence with other activities
- Any temporary works requirements
- The plant, small tools and equipment required
- The potential impact of the weather on the activity
- Quality of workmanship and materials required
- Wastage of materials and removal of the waste from site
- Delivery and storage of the materials
- Handling the materials to the point of use
- Supervision required for the activity
- The risk factors involved with undertaking the activity.

Understanding and bringing together these factors requires skill, knowledge, experience, data, information, assumptions and feedback on past performance. The estimating process requires systems and the ability to deal with complexity.

Computer-assisted estimating systems have helped and continually improve the estimating process with automated measurement, electronic tendering and the use of databases of information that can be manipulated for a particular project.

2.3 Profitability

Profitability is important for any organisation. Profitability is the ability of a business to earn a profit from its activities. A profit is what is left from the revenue after all the direct and indirect operating expenses have been taken into account. The challenge for construction is how to measure profit from complex tasks. Unlike a production line, construction projects are unique; they are undertaken by temporary project teams in variable conditions. The profit margin is often used as a simple measure of profitability, but it should take account of the return on capital employed in order to generate the revenue.

The profit margin should reflect the firm's ability to generate revenues and to control the costs of production. The construction sector has profit margins that should reflect the risk involved.

2.4 Types of tender

Open

For an open tender, the project is advertised publically in the press and on the World Wide Web; any organisation can apply for the documents and to bid, provided they meet the selection requirements. A fee may be charged to obtain the documents. The advantage is the wide range of companies likely to be interested in bidding. The disadvantage is that open tendering is indiscriminate and likely to attract bidders who are not suitably qualified for project. The cost of open tendering for the construction industry is very high.

On larger projects, there may be a pre-qualification process that produces a shortlist of suitable organisations who will be invited to prepare tenders. Open tendering has been criticised for attracting tenders/expressions of interest from large numbers of companies some of whom may be entirely unsuitable for the contract, and as a result it can waste a great deal of time, effort and money. However, open tendering offers the greatest competition and has the advantage of allowing new or emerging organisations to try to secure work.

Selected

In selective tendering, a selected number of suitably qualified contractors are invited to submit a bid based upon pre-qualification. The number of bidders should not be excessive, normally no more than five for a major project. Bidding consumes a lot of resources, time and money for the contractors and all the organisations in the supply chain; the decision to bid is not taken lightly as the cost involved can be substantial.

A framework agreement as 'an agreement or other arrangement between one or more contracting authorities and one or more economic operators which establishes the terms (in particular the terms as to price and, where appropriate, quantity) under which the economic operator will enter into one or more contracts with a contracting authority in the period during which the framework agreement applies.'

Negotiated

In negotiated agreements, a single general contractor (and sometimes a team of multiple contractors) prices the work as the sole bidder. Trust is an important part of negotiation with the contractor; frequently an owner has developed an ongoing relationship with a company. Negotiation is involved in a two-stage tender process where the successful contractor from the first stage negotiates the second stage of the bidding process.

Serial tendering

This approach can be used by the client when they have a number of similar projects. The pricing of the first project can lead to a number of sequential projects. This type of tendering can lead to economies of scale and efficiencies. The contractor has familiarity with the client's requirements and the design issues.

Framework tendering

This involves invitations to tender to contractors that operate a framework agreement with the client. The framework normally is for a fixed time period. A number of contractors can be selected for the framework.

A framework agreement is a general term for agreements with providers that set out terms and conditions under which specific purchases (call-offs) can be made throughout the term of the agreement. Framework agreements can be with a single provider or selected groups of providers. Framework projects are often agreed on the value for money (vfm) principle, rather than the lowest bid price.

2.5 Tender award

A contract must be awarded on the basis of either:

Lowest price: The lowest priced tender wins. No other element of the tender may be taken into account; or

The most economically advantageous tender (MEAT): Factors other than, or in addition to, price, like quality, technical merit and running costs can be taken into account. In MEAT, the contract award criteria, for example price, quality of services, risk to contracting authority and any sub-criteria must be set out in the tender documents, and the weighting of each criterion, and sub-criterion if weighted, must also normally be given, either as an exact number or as a meaningful range (e.g. 'price: 30–40%').

The evaluation process must be clearly stated in the tender documents and the extent to which they are mandatory, desirable or optional parts of the requirements. The objectives of this method are to identify the tender that offers best value for money, to compare bids on commercial, technical and financial terms, that is not just on price element and to allow for optional requirements to be priced and evaluated on a lifetime cost basis, that is, allows each criteria to be shown as either 'mandatory' or 'optional'. The estimator must therefore balance the bid to reflect the most advantageous offer.

Under English Law, a bid may be withdrawn any time before it has been accepted. Once a contract has been awarded, all the unsuccessful bidders should be informed to release them from any obligations on their resources. The tender will remain open for acceptance, normally for 60 days following the submission date. The client does not have to accept the lowest or any tender submitted.

2.6 The difference between procurement and tendering

Procurement is the overall act of obtaining goods and services from external sources (i.e. a building contractor) and includes deciding the strategy to solicit tender offers, how those goods are to be acquired by reviewing the client's requirements (i.e. time, quality and cost) and how the tenders will be evaluated, the attitude to risk and how the contract will be awarded.

In summary, procurement is about:

- establishing what is to be procured
- deciding on the procurement strategy
- soliciting tender offers
- evaluating tender offers based upon the defined selection criteria
- awarding the contract.

Tendering is an important phase in the procurement strategy involving more than simply obtaining a price. Tendering is:

- the bidding process to obtain a price and duration
- appointment of the successful contractor following bid submission.

2.7 Methods of procurement

The estimating/bid team face a wide array of different procurement methods. Successful procurement relies upon all parties complying with their respective obligations and managing risks appropriately. Contractors bid to win work, not just by being the lowest bidder on price. Some clients select on lowest price, health and safety record, environmental sustainability, quality record, innovation, ethical responsibility or confidence/trust in the individuals likely to work on the project.

New methods of procurement are constantly evolving, with new forms of contract embodying different approaches to the management of risk – see Selection, **Procurement, contractual arrangements and legal issues chapter.**

Cost implications at the bid stage

Methods of procurement have cost implications for the bidding process. Projects won on a whole-life basis have long-term implications, with the commitment to operate and maintain the project over a time horizon into the future. Submissions based on best value, rather than lowest price, require detailed tender submissions beyond priced bill quantities, schedule of rates or priced specification and drawings.

Price is not the only determinant

Quality will always secure projects. Quality assurance, quality control and quality management are embedded in construction practice with quality marks, ISO 9000 certification, total quality management systems and industry good practice.

Quality assurance is a way of preventing mistakes or defects in manufactured products and avoiding problems when delivering solutions or services to customers, which ISO 9000 defines as part of quality management, focused on providing confidence that quality requirements will be fulfilled. Quality control makes sure that the results of what was done are what was expected. The requirement for quality is embedded in the construction process; however, bid prices must reflect the quality the client is expecting.

3 Changes driving estimating practice

Construction is changing to reflect the need for a safer, higher quality, faster and friendlier industry that has safety and health, quality, sustainability, public responsibility and ethical behaviour, at its core. Clients* do not always want faster delivery; they want certainty with more reliable cost and time delivery to the highest quality, with no surprises, and a project fit for purpose. From strategic definition through to occupation, the client is concerned with the price for the project to ensure that it meets the functionality, performance and quality requirements.

Digitisation has cost implications, as does digital security with the need to keep data safe. Digital security includes the tools used to secure identity, assets and technology in the online and mobile world.

Winning work has become more costly, more complex and involves more parties.

Rules and regulations for all aspects of construction, including safety and health, sustainability, environmental protection, energy efficiency and corporate social responsibility, have increased. Contractors, specialty† contractors and suppliers must make allowance in their bids for the cost of compliance. Stringent regulatory requirements have also led to cost increases in other areas, such as temporary works and the cost of the removal and disposal of waste from sites. Insurances have increased in cost with the increase in claims and a claim culture in society.

3.1 Digitisation

The increasing use of digitisation through CAD drawings and the use of building information modelling (BIM) has changed both design and production methods. Modern 3D software allows for a rapid proof-of-concept, eliminating many of the errors encountered after using traditional design techniques. 3D modelling also paves the way for more sophisticated tasks to take place as well as enabling downstream data management in the delivery phase. New information and communication technologies have presented both challenges (information overload) and benefits (better communication, better connectivity). The estimator has greater access to information from

* The term client (individual, company or organisation), employer, owner, promoter are used interchangeably in documentation for the tendering process to mean the person or organisation acting as the body buying the goods and services. For consistency, the terms client and employer have been used throughout.

† The term specialty contractor has been adopted throughout the CoEP. In some countries, the term sub-contractor, works contractor, or trade contractor is used; the term co-contractor is also used. Specialty contractors can be micro, small, medium, and large companies that have a specialism that may involve bidding for supply only materials, labour only services, supply and fix, design supply and fix.

New Code of Estimating Practice, First Edition. The Chartered Institute of Building.

all stages of the process; dealing with that wealth of information is a significant challenge.

Bid management software helps to keep track electronically of the bid documents. Collaboration software ensures that all the parties share documents. Cloud computing† provides software as a service, allowing greater access to information stored in the public or private cloud.

Computer-assisted estimating systems allow contractors to use standard databases of prices or to customise price information. Computer planning and scheduling systems help in allocating resources to work activities. Costing systems can balance the cost against each of the estimated work packages.

3.2 E-Tendering

E-Tendering has become more widely used with an RICS e-tendering compliant guide. Online tenders are published as digitally sealed bids into an online safe deposit box. Specified representatives can only access them on the pre-determined tender closing date. Online tender management systems are frequently used to manage the bidding process.

3.3 Legislation and taxation

Legislation has increased for the industry with a focus on health and safety and environmental codes and standards. The Construction Design and Management (CDM) Regulations impose a duty of care on contractors, designers and clients to ensure that construction is planned, managed and monitored, so that construction work is undertaken safely and without risks to health. Environmental protection and sustainability have become important requirements, with the emphasis on recycling, waste management, energy efficiency and use of renewable materials. New taxes impose greater burdens on the construction sector, with the Carbon Tax§ and Landfill Tax 2011** being examples. Industry training levies are not new, but they are more wide-ranging in scope.

3.4 Bureaucracy

Increasing bureaucracy has added to the complexity of the estimator's work. The 2007 financial crisis spawned new legislation on ethics and corrupt practice, placing a burden on compliance that has affected cost estimation. All the stakeholders now have a bigger involvement in a project, with consultation and agreement reached on how to ensure all the parties are satisfied at the bid stage. Such stakeholders included the workforce, the supply chain and the general public.

3.5 Competition and winning work

Competition remains a fundamental requirement on most contracts for the supply chain. Competition for work has grown stronger and more complex, with some design, engineering and cost consultants prepared to undertake project delivery with fee-based construction management.

† Cloud computing is the practice of using a network of remote servers hosted on the Internet to store, manage and process data, instead of a local server or a personal computer. In cloud computing, everything is treated as a service, for example, SaaS (Software as a Service), PaaS (Platform as a Service) and IaaS (Infrastructure as a Service).

§ A carbon tax is usually defined as a tax based on greenhouse gas emissions (GHGs) generated from burning fuels. It puts a price on each tonne of GHG emitted.

** A landfill tax or levy is a form of tax that is applied in some countries to increase the cost of using landfill sites. The tax is typically levied in units of currency per unit of weight or volume (£/t, €/t, $/yard^3).

3.6 Specialisation

Specialisation is an important part of modern-day contracting. Contractors manage resources and speciality contractors and suppliers. Sub-contracting work packages to specialists is a fundamental part of project delivery for any contracting organisation. Micro-, small- and medium-sized as well as major contractors are all reliant upon specialists. Specialty contracting and estimating is more widely discussed later in this Code of Estimating Practice (CoEP).

3.7 New technologies and off-site production

New technologies in both design and production have influenced estimating practices. Advances in materials, components, plant and equipment as well as production processes all have an impact on cost estimation.

Off-site production, pre-fabrication of components, design for manufacture and assembly (DfMA) are all terms used frequently to reflect the desire for off-site assembly. Whether supplied as fully volumetric models, structural insulated panels, laminated timber framing systems or plumbing and electrical systems assembled off-site, the result is that the estimator will use specialist companies that will price their part of the work based on supply only or supply and install. The estimating process requires a full understanding of all the interfaces and specialisms to incorporate such systems into the project.

The CoEP recognises the changing nature of construction; it provides a basic set of procedures that represent best practice from a commercial and professional point of view, as a practical guide for practitioners and for students and academics. It provides an overview of the systems involved and the underlying principles that guide bid managers and estimators to establish and maintain sustainable, sound and efficient systems.

3.8 New methods of procurement

The need for good estimating arises from the most common form of procurement: competitive lump sum fixed price tendering. At some point in the supply chain, lump sum competitive tendering is required to enable the selection of contractors or speciality contractors to carry out work packages.

Despite attempts to move towards a less confrontational approach, single-stage competitive tendering remains many clients' favoured route to secure best value. Fee-based methods, such as management contracting and construction management, move the major focus of competitive tendering 'down' the supply chain from principal contractor to works package contractor selection.

The private finance initiative (PFI) and public private partnership (PPP) methods move the first stage of competitive tendering 'up' the supply chain to the design, build and operate developer and will usually require that estimators take much greater cognisance of whole-life costs, as opposed to purely capital construction cost.

The growth of target cost/guaranteed maximum price contracts places additional demands on the estimating and tender settlement functions.

3.9 Best value

The construction industry is under pressure to innovate and make on-site production more like a manufacturing process. Innovation can involve the contractor offering an alternative option that may save the client money and time on the project. Innovation involves finding ways to ensure that the production process can be efficient, safe and beneficial to all the parties.

There is a welcome change in procurement practice to move away from lowest price to best value as the leading selection criterion. The result of pre-construction practice has included an increase in demand for contractor input into the following:

- Design and buildability/constructability
- Design for manufacture and assembly (DfMA)
- Design management
- Design completion††
- Alternative option procurement
- Demonstration of best value
- Whole-life cost assessment
- Value engineering.‡‡

Proposals are often required as a part of detailed bid submissions, giving clients the means to select contractors on more than lowest price. This has changed the outputs of the bid process from the simple 'form of tender' to full-blown contractor's submissions incorporating design, planning and logistical proposals along with detailed costings. This places far greater demands on a contractor's estimating process in terms of the quantity and quality of inputs required.

The importance of the supply chain and the speciality contractors, material suppliers and merchants are vital for any project. Specialty contractors are increasingly relevant in the demonstration of best value; it is their resources that help to realise the benefits gained from innovation and technological development.

The need for assurance of high-quality delivery from speciality contractors encourages contractors to enter into formal and informal partnering arrangements with specialist contractors who can demonstrate a good record of accomplishment. This changes the procurement approach for some contractors and some tenders, making the cost negotiations of work packages a preferred route, with assurance of best value.

The estimator's relationship with specialty contractors during the tender period must be much closer, with more trust and collaborative working. Working in this way imposes some restrictions on the way in which some estimators have, in the past, managed the process of procuring competitive sub-contract quotations.

†† A design completion service enhances the traditional design process by taking 2D or 3D designs and completing them in a 3D environment.

‡‡ Value engineering promotes the substitution of materials and methods with less expensive alternatives, without sacrificing functionality. It is focused solely on the functions of various components and materials, rather than their physical attributes. It is also called value analysis.

4 The bid process

4.1 Work breakdown structure

Modern construction is about breaking a project into a work breakdown structure with bid packages undertaken either by the general contractor with directly employed labour or by specialty contractors. The specialty contractor can provide a labour only or labour and material price. Some specialist suppliers will provide a supply only or a supply and fix price.

Figure 4.1 shows the estimating 'honeycomb' with a breakdown of a project by elements and sub-elements in each hexagon; the elements may be a bid/work package in their own right, or each element may have more than one work package. The Code of Estimating Practice (CoEP) further sub-divides these to ensure that both cost and non-cost considerations are considered. The honeycomb graphic helps to show interrelationships in a simple way.

The honeycomb structure shown in Fig. 4.1 is based upon the new rules of measurement (NRM) published by the Royal Institution of Chartered Surveyors. The concept is equally applicable for any rules of measurement used to measure work.

The NRM2 provides fundamental guidance on the detailed measurement and description of building works for the purpose of obtaining a tender price. The rules address all aspects of bill of quantities (BQ) production, including setting out the information required from the employer and other construction consultants to enable a BQ to be prepared, as well as dealing with the quantification of non-measurable work items, contractor designed works and risks. Guidance is provided on the content, structure and format of BQ as well as the benefits and uses of BQ. While written mainly for the preparation of a BQ, quantified schedules of works and quantified work schedules, the rules are invaluable when designing and developing standard or bespoke schedules of rates.

New Code of Estimating Practice, First Edition. The Chartered Institute of Building.

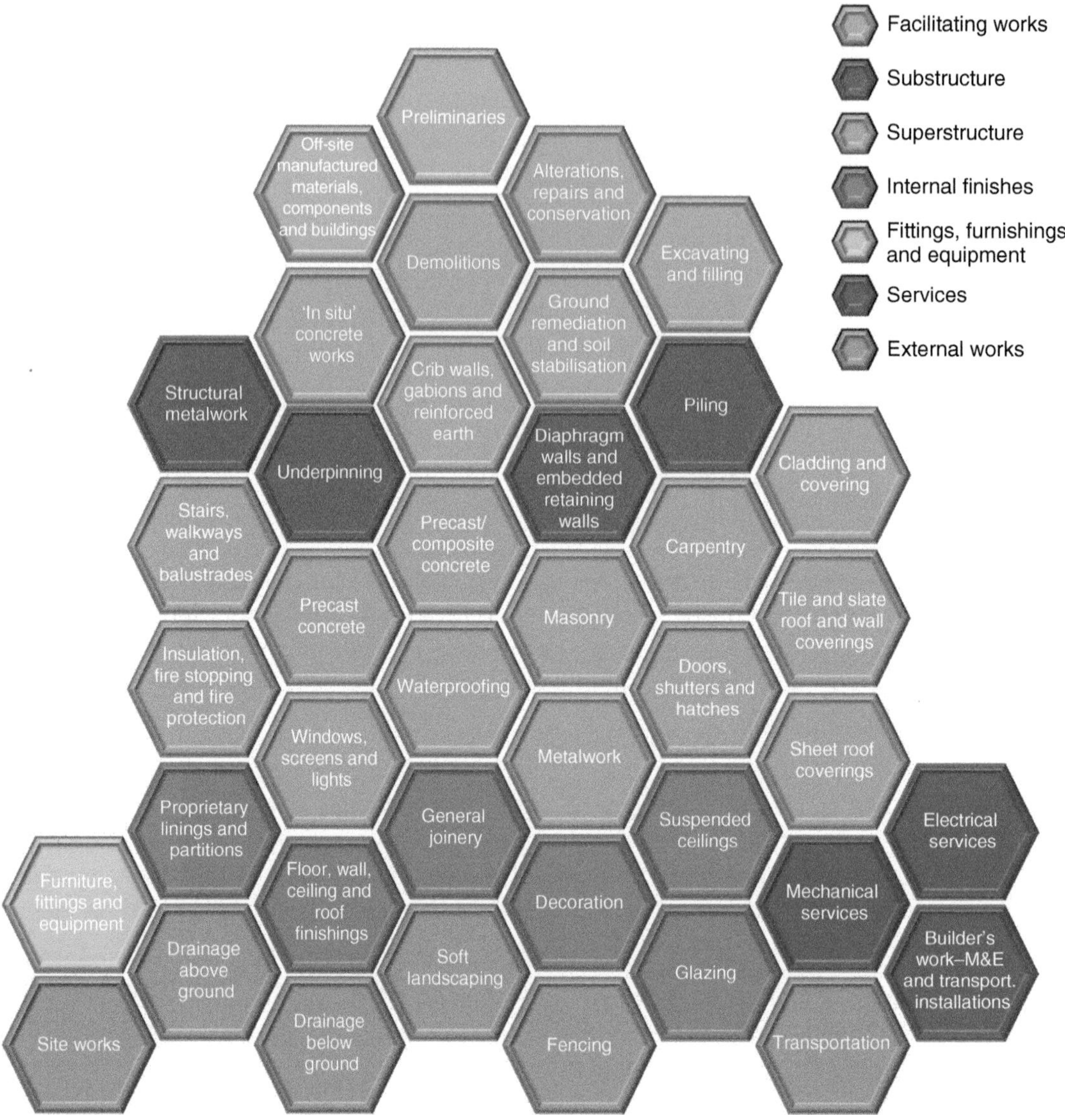

Figure 4.1 The estimating honeycomb.

4.2 Underlying principles

General good practice

The client's needs

Establishing what the client wants from a project should be of prime importance. The tender submission document that accompanies a bid should demonstrate how these needs are addressed.

Professionalism

Good practice involves acting in a professional manner in all dealings with clients and consultants, specialty contractors and suppliers. Such dealings should be honest, straightforward, thorough, accurate and informed. Communications should be capable of being understood, appreciated and acted on. Professionalism adds authority to what is done and makes activities more effective and successful.

Investigations

Where a design is provided with tender documentation, a design review is essential in order to arrive at a clear understanding of what is required. A query list should be produced at the outset and seek to have meetings with the design team to get their explanation of the project and responses to queries (should the nature of the tender and the timescales permit). By being aware of the options available, a contractor can seek a competitive edge with better value (alternative) products or innovative techniques. Information gained during the tender stage can be essential to winning the tender and ultimately to the successful completion of the project.

Error avoidance

One of the basic principles of estimating is error avoidance; estimates cannot be guaranteed to be error free, but good procedures will keep errors and their impacts to the minimum. If a tender is very low due to errors, it may be disregarded by the client, or alternatively, it may be accepted, and the contractor could make a substantial loss. Clearly set out procedures are essential when working as part of a team as they enable others to undertake work, to complete it or take it on to the next stage. Logical, transparent procedures enable estimators to explain and demonstrate what is required to new colleagues. When there is a successful tender, the construction team would expect the estimator to hand over material and workings that are clearly recorded. The information used must be readily available with all accompanying forms, analyses and reports.

Leadership

Traditionally, estimators have led the bidding team at the tender stage, setting the agenda and facilitating the process. With the increasing demands for more sophisticated and detailed tender submissions, there has been the increasing use of a bid manager.

Tender requirements have become much broader and often require a multi-skilled team to carry out the bidding process. On larger tenders, bid managers or even project managers will lead the tender team. On small or more straightforward tenders, the estimator may lead the process, requiring skills in communication, team working and co-ordination. Estimators should see it as part of their professional duty to take on a leadership role and to progress to becoming bid managers.

Company policies

A tender is a particular company's solution to the needs of a particular project and, as such, must present that company's policies in all elements of the tender, from the method statements in the tender submission to the preliminaries, overheads and profit. If a tender does not comply with company policies, the construction team may not be able to complete the contract successfully.

Estimators must keep themselves and their supply chain informed of the relevant aspects of company policies to make sure that they are incorporated into the tender costs. It is equally important to make the client aware of policies, where they are relevant and where they meet a client's needs and could add value to the tender. An estimator should ensure that the pricing of the preliminaries includes all necessary costs arising from the implementation of company policies, for instance on health, safety, welfare, ethical and environmental issues and be able to demonstrate how the contract could help to meet the objectives of the company's business plan in terms of workload and profitability.

Contractor selection

The increasing sophistication of construction clients and the complexity of their needs have led to the selection criteria for contractors on best value principles, rather than lowest cost. Issues such as the proposed project delivery team, construction method, project sequence and duration and value addition can be more important than cost in selecting the winning tender.

Clients sometimes use selection panels, interviews and scoring matrices to assist in contractor selection. Where a collaborative form of procurement is used, it is likely that significant emphasis will be placed on an assessment of the extent to which the contractor's business ethics and core management philosophies match those of the client. Presentation of the tender submission documents is crucial. In order to demonstrate that a particular tender provides the best value, it must adequately express the crucial issues in the proposal documents in written and drawn method statements, diagrams and design proposals. At a post-tender interview, it can be beneficial to utilise the latest technology alongside good presentation skills to communicate a proposal.

Clients and consultants are usually well practised in interpreting tender submissions and scoring tender interviews. They will not be impressed by volumes of standard information taken off the shelf and reproduced in a tender submission simply to fill up space. Proposals that meet the needs of the client and showing professionalism in presentation are important. Slick interview teams who will never play any further part in their project are unlikely to make a good impression; the team should be comprised of the people who will be working on the contract, and they need to demonstrate how they will approach the key project challenges.

The competitive edge

The cost of tendering rises in proportion to the increases in the requirements of tender submissions; it becomes all the more important for a contractor to maintain an adequate success rate when tendering. The estimator should aim to achieve the highest winning bid while maximising profit levels. This can only be done when the lowest, accurate and reliable estimated cost for the project is known.

Bid success rates will vary with the type of work and the size of projects. A rule of thumb is that contractors have a 1–4 or 5 success rate in winning lump sum traditional tenders. Across the construction sector, it means that contractors, specialty contractors and suppliers must absorb the real cost of bidding into the overhead costs.

Effective procurement of specialty contract works/works packages is vital, often making up over 80% of the construction work. The estimator should look for a competitive advantage, through alternative more economical suppliers, design simplifications (although this may incur design liabilities), innovative approaches to production and more cost-effective methods of delivering the project.

Winning tenders

It is sometimes argued that the level of profit and overheads is the main influence on winning the project; this is only partly true. The key is finding the most cost-effective and innovative production process, combined with an efficient purchasing approach to buying materials and works packages. Through careful interpretation of the client's requirements, an estimator can submit a very competitive tender with alternatives for the client's consideration.

Allowances must be made for the likely impact of inflation on wages, transport and material prices through the project. The usual process is to use judgement based upon experience and cost indices. Inevitably, it is not scientific, but on a 2- or 3-year project, some allowance has to be built into the bid price in any fixed price bid.

Qualifying the tender

Qualifications, or tender notes, can be used to set aside items that the estimator believes should be excluded and to limit risks, for instance by giving provisional sums for unclear items rather than adding a risk allowance to the tender (should the tender document allow). Such strategies, if clearly explained, are legitimate and, indeed, often necessary in competitive tendering. The risk and implications of a non-compliant tender should also be considered at the tender settlement stage by senior management.

Design completeness

It is false economy to start a project with poor information. Incomplete, inadequate and poorly detailed design information at the tender stage costs money for the contractor and ultimately the client - the risk allowances that must be added to the bid may not be correct.

The level of design completion at the tender stage is an important factor. The more complete the design, the fewer the unknown factors. The pricing of services to be procured from specialty contractors can create several iterations, especially if not all the information is available at the appropriate time. No specialty contractor is going to give a fixed price bid on incomplete information; their price will be qualified to reflect the unknown items, resulting in the contractor having to decide what risk allowance should be added to reflect the unquantified risk in the main bid.

In practice, tenders are frequently sought on designs that are only 50–60% design complete through no fault of the design team. Pressure is placed on the team to seek tenders in order to start the project as quickly as possible. Ambiguity and unknowns will add to the risk and to the price; it is a false economy to start construction work too soon with incomplete information under a traditional design–bid–build approach (competitive tendering without any design input). Variation and change orders issued during construction disrupt the construction process, costing the contractor money because of uneconomical working.

Bid management

The traditional lump sum tendering process was straightforward. However, in recent years, other forms of procurement have become more prevalent, including design and build (D&B), engineer procure construct (EPC), early contractor involvement (ECI), partnering, alliancing and so on. Procurement approaches have increased; the bid team has become more specialised, the process more complex, more time consuming and requiring more resources. Ultimately, with the need for increased bid management.

For a large D&B project, the bidding team may include:

- Architects
- Structural engineers/civil engineers
- Mechanical, electrical and plumbing services engineers

in addition to the contractor's own:

- Estimator
- Planning engineer/project scheduler

- Construction/production manager
- Design manager
- Purchasing and supply chain manager
- Plant and equipment manager
- Building information modelling (BIM) manager.

In the estimating department nobody can afford to miss a tender submission deadline. Failure means a waste of cost, time and an opportunity to win new orders.

Managing the bid process is critical in compiling the bid. In a perfect world, people are working on one project at a time; in reality, there are overlapping deadlines with many projects being worked upon. That makes bid management even more important.

For the small- and medium-sized enterprise (SME), the problem is just as critical; without winning new orders, the company cannot survive. The SME will not have the luxury of specialist staff to prepare the bid. Often, the principals will be multi-tasking, engaged in marketing, planning, estimating, procuring materials and the production process. Estimating has to be undertaken alongside all the other tasks.

Supply chain management at the bid stage

For clients and consultants, the estimator will be introducing specialty contractors and suppliers who will be critical to the project success. Larger contractors will have a buying or a procurement department or supply chain manager who may take the prime role in developing and maintaining relationships with suppliers and speciality contractors.

Most medium and large contractors keep an approved list of specialists when procuring services, which records information about performance on-site, financial and insurance details, quality assurance and health and safety compliance. Such lists are critical to effective maintenance of performance quality and compliance to legal and policy criteria. Some projects require cost-effective and quick solutions, whilst others require innovation or high-quality thinking from the supply chain. Keeping this type of information helps in using the appropriate specialist on a specific project.

Value management *A systematic and organised approach to provide the necessary functions in a project at the lowest cost. Value engineering promotes the substitution of materials and methods with less expensive alternatives, without sacrificing functionality. It is focused solely on the functions of various components and materials, rather than their physical attributes.*

Effective supply chain management that can be demonstrated to clients is an excellent contribution to the added value that a contractor can bring to a project. It will give clients an assurance of the performance that a contractor will provide, particularly where they have not worked together before.

Value management/value engineering/value analysis

Value management is a systematic and structured approach for improving projects, products and processes. Also known as value engineering and value analysis, it is used to analyse and improve products and processes for construction projects.

Value engineering is of most use when estimating for a D&B tender or for a two-stage tender as a demonstration of value for money. Value management gives the client assurance that they are getting the best value on the project, while maintaining the design, quality, programme and cost objectives.

The contractor may be required to value engineer a project that is over budget, by identifying areas with the design team where a more cost-effective design solution can be used. It is not simply cost cutting; it has to be based upon solutions that can be delivered.

Project planning/pre-tender programme

All projects need planning to identify the sequence and relationship of work packages and activities. Logistical solutions need to be found, which may be simple or complex; depending on the project, they may be met by the contractor's own resources or by outsourcing. The project requirements and alternative solutions must be considered early on in the tender process in order to plan the project duration, sequence and to cost the necessary resources.

It is not feasible on small- and medium-sized projects to use extensive planning until the project is secured. A bar chart will show key dates and a sequence of work. On larger projects, the pre-tender programme reflects the durations and the relationships between work packages.

More information about project planning is in Resource and Production Planning in the Principles section.

5 Health and safety

5.1 Introduction

Health and safety is non-negotiable; it is of paramount importance to everyone in construction.

Health and safety issues must be considered at a very early stage to ensure compliance and the highest possible safety record. These issues need to be identified, planned, organised, controlled, monitored and reviewed. The responsibility for health and safety rests upon all those controlling site work. Planning needs to take into consideration any changes that may occur as the project develops, from welfare arrangements at the set up, through to snagging work and the dismantling of site huts and hoardings at the end of the contract.

New health and safety risks need to be considered, such as the off-site pre-fabrication of complete bathrooms. These include the construction of a skeletal structure on-site and the subsequent installation of the bathrooms as well as new substances, such as adhesives and surface finishes, that provide increased in-use performance, but which may have added occupational safety and health risks during construction. The risks of new machinery and plant for gaining worker access at greater height need to be taken into account.

There are rapidly changing legal and moral obligations that need to be fulfilled with regards to health, safety, welfare and environmental (HSW&E) issues. There is a body of law on health and safety and corporate responsibility that must be adhered to. Operatives' welfare similarly is covered by law but is also a practical employment issue since poor welfare facilities discourage and demotivate operatives and specialty contractors, leading to lower productivity. Whether obliged by law or policy, the cost of appropriate HSW&E issues must be addressed when preparing an estimate. The key HSW&E issues must be addressed, including those imposed by (HSE, 2006; HSE 2015b) CDM (2015) regulations.

5.2 Estimating the cost of health and safety

Health and safety is non-negotiable; it is of paramount importance to every company operating in the construction sector. The cost of health and safety across the duration of a project should be taken into account at the estimating stage. A health and safety programme should ensure the safety of the public, users and operators, and meet legislative requirements. Health and safety costs can be divided into three cost categories: prevention, insurance and accidents. Table 5.1 shows the different parts of each category.

New Code of Estimating Practice, First Edition. The Chartered Institute of Building.

Table 5.1 Categories of health and safety costs and their components.

Category	Component
Insurance	■ Personal (occupational injuries, illnesses and fatalities) ■ Company (public liability as a result of any accident) ■ Professional indemnity insurance to cover the cost of any design liability for health and safety ■ Checking insurances of the specialty contractors to ensure they meet the minimum requirements for cover
Prevention	■ Protection of individual workers (PPE) ■ Protection for all site operatives and managers ■ Safety systems for plant, tools and equipment ■ Safety systems for storage, transport and management of materials ■ First aid installations, equipment and supplies, including the cost of any medical staff that may be employed on a large project ■ Administration and management ■ Developing the health and safety management plan ■ Reviewing, updating and revising the health and safety file ■ Monitoring and coordination of the health and safety plan with workers and specialty contractors ■ Training and induction on health and safety
Accident	■ Direct (approx. 25% of the cost of an accident) Medical expenses Compensation Cost of any fines imposed by the authorities ■ Indirect (approx. 75% of the cost of an accident) Production time lost due to an accident – injured worker and the associated workforce, possible temporary closure of project during an investigation Cost of hiring and training a replacement worker Programme interruption and process delays to the project Accident management Accident investigation Legal fees Loss of business by impacting the likelihood of winning further work by having a poor accident recorded that must be stated on any pre-qualification document. Loss of reputation

It is difficult to allocate any particular cost item above to a single activity; the cost of the health and safety requirement will generally be included within the general preliminaries costs for the overall project. Health and safety costs are not captured by a costing system; some costs are embodied within the head office and site overhead items.

Personal Protective Equipment (PPE)

The regulations covering PPE are the Personal Protective Equipment at Work Regulations (1992).

Personal protective equipment (PPE) is all equipment, which includes clothing that is intended to be worn by the operative. It protects them against the risks to health and safety. PPE includes safety helmets, gloves, eye protection (safety spectacles, visors, goggles and face shields), safety footwear, safety harnesses and high-visibility clothing. In certain circumstances, hearing protection may be necessary, as may respiratory protective equipment. PPE must be supplied whenever there are risks to health and safety. It must be properly inspected before use, maintained and stored properly and used correctly, with instruction on proper use.

Anyone using PPE needs to be aware of why it is needed, and when it is to be used, necessitating training and induction.

Safety signage around the site provides greater awareness for the operative. On almost all construction sites, the risks of head injury are such that the law required head protection to be used. Some construction operations, such as temporary traffic management workers, need a higher standard of visibility. The Health and Safety (First Aid) Regulations 1981 require all construction sites to have:

- A first aid box with enough equipment to cope with the number of workers on-site
- An appointed person to take charge of first aid arrangements
- Information telling the operatives the name of the appointed first aider.

The requirement is that there should be one trained first aider for every 50 people employed on-site.

Construction (Design and Management) Regulations – CDM Regulations

The CDM Regulations were introduced in 1994, following publication of the 1992 European Directive on the minimum standards for health and safety on construction sites. They had a substantial rewrite in 2007 aiming to simplify the construction process, improve coordination between the parties and to reduce bureaucratic burden. The latest version was released in 2015. Figure 5.1 shows the regulations within CDM (2015) broken down into the honeycomb, all of which have a cost impact or need to be taken into account where applicable.

The aim of the changes to CDM (2007) was to simplify the regulations, re-align them with EU directives, expand the definition of a client and to change when key duty hold-

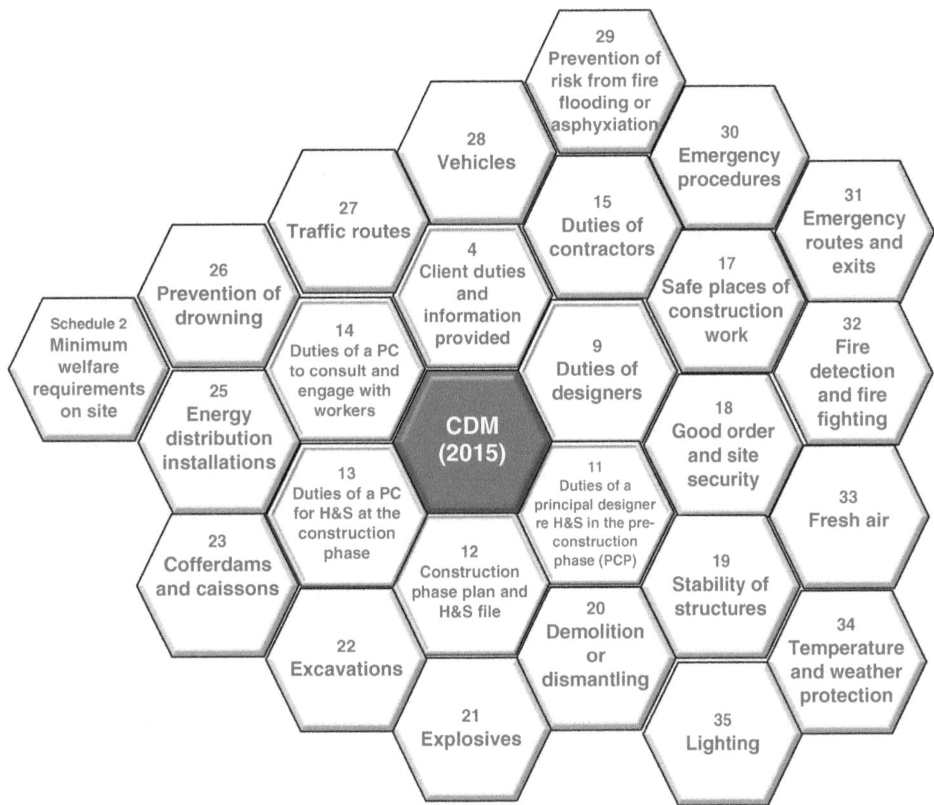

Figure 5.1 CDM (2015) regulations.

ers are appointed. There are a number of major changes between the two sets of regulations:

Principal designer: The replacement of the CDM co-ordinator role (under CDM, 2007) by principal designer means that the responsibility for coordination of the pre-construction phase – which is crucial to the management of any successful construction project – will rest with an existing member of the design team.

Client: The new Regulations recognise the influence and importance of the client as the head of the supply chain, and that they are best placed to set standards throughout a project.

Competence: This is split into its component parts of skills, knowledge, training and experience, and, if it relates to an organisation, organisational capability. This will provide clarity and help the industry to both assess and demonstrate that construction project teams have the right attributes to deliver a healthy and safe project.

Figure 5.2 shows the hierarchy of responsibilities in CDM (2015), where designers only have implied duties. The responsibility of taking account of health and safety implications of the design now rests with the client or the project supervisor.

European Directive 92/57/EEC – the implementation of minimum safety and health requirements at temporary or mobile construction sites

The Construction Sites Directive (92/57/EEC (5)) lays down minimum safety and health requirements for all temporary or mobile construction sites, irrespective of their size and complexity. The Directive has brought about major changes in the area of occupational risk prevention in the construction sector by:

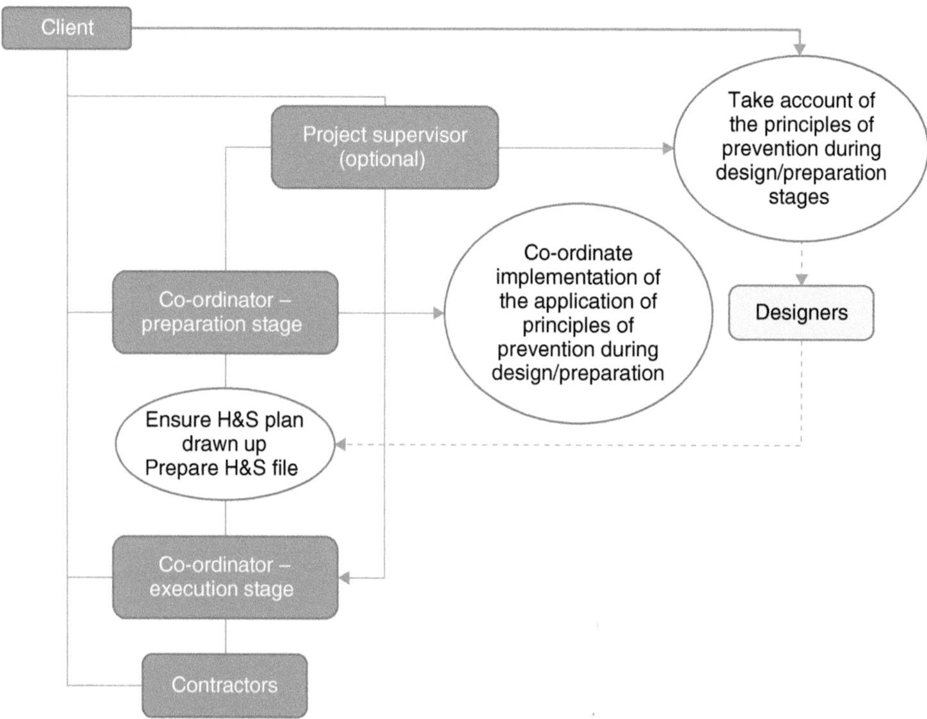

Figure 5.2 The hierarchy in CDM (2015). *Source:* CDM (2015).

- Requiring the preparation of a limited number of documents that assist in ensuring good working conditions;
- Making clear the roles and responsibilities of the various stakeholders;
- Extending to all of the players involved in construction projects the principles that are found in the Framework Directive for undertakings sharing a workplace to co-operate and co-ordinate in preventing occupational risks;
- Requiring safety and health coordination for both the project preparation stage and during project execution stages.

Health and safety is discussed further within the Processes section.

6 The pre-qualification process

6.1 Introduction

Pre-qualification is a necessity for public sector and non-governmental organisations (including central government authorities, local authorities, universities, NHS trusts and some utility companies) and with some large private sector clients with an ongoing programme of work. It is used for the establishment by the client of a list of contractors or specialty contractors with the necessary skills, experience, resources, previous tender performance and desire to carry out the works, bearing in mind the character, size, location and timing of the project.

Pre-qualification will occur some time before any formal invitations to tender are issued. It is important for the client to allow a realistic programme covering the whole of the pre-qualification and tendering period, and that adequate time is given for each stage.

Comprehensive and detailed information will be required; on major works, such submissions can take several weeks to prepare. The European Union (EU) has requirements regarding time periods when tendering for public sector works.

Pre-qualification procedures can be used on a project-by-project basis or for a framework agreement. Being pre-qualified does not guarantee that the contractor will be chosen for the tender list. Excessive pre-qualification, with long burdensome questions, adds cost, bureaucracy and frustration to the tendering process.

*BSI PAS*91: 2013*

BSI PAS 91 is a publicly available specification that sets out the content, format and use of questions that are widely applicable to pre-qualification for construction tendering. It states that 'To be eligible for pre-qualification, it is necessary that suppliers demonstrate that they possess or have access to the governance, qualifications and references, expertise, competence, health and safety/environmental/financial and other essential capabilities necessary for them to undertake work and deliver services for potential buyers. ... The use of this set of common criteria by those who provide pre-qualification services will help to streamline tendering processes.'

Having to submit frequent pre-qualification questionnaire (PQQ) forms is expensive and time-consuming. Many of the questionnaires are in different formats, and the

* PAS – publicly available specification.

New Code of Estimating Practice, First Edition. The Chartered Institute of Building.

Table 6.1 Example of a checklist for PAS 91 process – to be compiled by the bid team.

Project	Location
Client	Previous projects with client
PAS 91 return date	

PAS 91 Module	Information required	Person/Department responsible	Information requested (date)
Core questions			
1	Supplier		
2	Key roles and contact information		
3	Financial capability		
4	Company ethos/core competencies		
6	Health and safety record		
Optional questions			
1	Equal opportunity policy		
2	Environmental management		
3	Quality management		
4	BIM capability		

variety of forms can drain resources. Hence, BSI PAS 91 provides a consistent and uniform basis for pre-qualification. It also specifies the requirements for the consistent use of those questions across projects of varying sizes and types, including the OJEU procurement thresholds for public sector procurement.

BSI PAS 91 specifies what is to be asked in the pre-qualification process for construction-related procurement. It does not stipulate how the enquiry process is to be undertaken. Its aim is to underpin good practice.

The pre-qualification process has become an important part of tendering, with the need to demonstrate competence and capability to deliver projects.

Pubic sector buyers are often committed to promoting certain policy objectives, such as local employment, apprenticeships, training, environmental requirements and use of small- and medium-sized enterprises (SMEs). Such items may be covered in the questionnaire. The purpose of the questionnaire modules – see Table 6.1 – Is to extract information about the organisation seeking to be qualified to bid.

6.2 Bidding for public sector projects

Bidding on public sector projects in member countries of the European Union is governed by EU Directive 2014/24/EU. This legislation lays down the rules for awarding contracts for public works, supplies and services. It aims to ensure that the contracting process is fair and open to bidders from anywhere in the EU. In the United Kingdom, this Directive is implemented by the Public Contracts Regulations 2015.

The Directive is an update on the 2004 version and is designed to facilitate SME participation by:

- Encouraging contracting authorities to break contracts into lots to facilitate SME participation.

- Introducing a turnover cap – contracting authorities will not be able to set company turnover requirements at more than two times contract value except where there is a specific justification.
- Helping suppliers to bid cross-border by providing a central, online point called 'E-certis' where suppliers can find out the type of documents, certificates and so on which they may be asked to provide in any EU country, even before they decide to bid.

The 2014 version of the Directive provides a much simpler process of assessing bidders' credentials, involving greater use of supplier self-declarations, and where only the winning bidder should have to submit various certificates and documents to prove their status. Other procedure changes include:

- Preliminary market consultations between contracting authorities and suppliers are encouraged, which should facilitate better specifications, better outcomes and shorter procurement times.
- More freedom to negotiate. Constraints on using the competitive procedure with negotiation have been relaxed, so that the procedure will generally be available for any requirements that go beyond 'off-the-shelf' purchasing.
- The statutory minimum time limits by which suppliers have to respond to advertised procurements and submit tender documents have been reduced by about a third.
- Full electronic communication (with some exceptions) become mandatory for public contracts 4.5years after the Public Contracts Directive came into force (i.e. October 2018).

The 2014 version of the Directive has improved rules on social and environmental aspects, making it clear that:

- Social aspects can now also be taken into account in certain circumstances (in addition to environmental aspects which have previously been allowed).
- Contracting authorities can require certification/labels or other equivalent evidence of social/environmental characteristics, further facilitating procurement of contracts with social/environmental objectives.
- Contracting authorities can refer to factors directly linked to the production process.
- The full life-cycle costing can be taken into account when awarding contracts; this could encourage more sustainable and/or better value procurements which might save money over the long term despite appearing on initial examination to be more costly.
- Legal clarity that contracting authorities can take into account the relevant skills and experience of individuals at the award stage where relevant (e.g. for consultants and architects).

Contracting authorities may make known their intentions of planned procurements through the publication of a prior information notice (PIN). The purpose of this is to inform the market of impending work; it allows the use of reduced timescales. A PIN could be used for services or supplies – published at the start of the financial year. For works, the PIN would be published once a decision has been made to proceed with a works contract, for example funding agreed.

The Directive covers most public contracts other than for utilities (water, transport, energy and postal services), telecommunications, service concessions (such as operating an existing car park) and certain defence and security contracts. It has five award procedures:

- *The open procedure*, under which all those interested may respond to the advertisement in the OJEU by submitting a tender for the contract.
- *The restricted procedure*, under which a selection is made of those who respond to the advertisement, and only they are invited to submit a tender for the contract.
- *The competitive dialogue procedure*, under which a selection is made of those who respond to the advertisement, and the contracting authority enters into dialogue with potential bidders, to develop one or more suitable solutions for its requirements and on which chosen bidders will be invited to tender.
- *The competitive procedure with negotiation*, under which a selection is made of those who respond to the advertisement and only they are invited to submit an initial tender for the contract. The contracting authority may then open negotiations with the tenderers to seek improved offers.
- *The innovation partnership procedure*, where a selection is made of those who respond to the advertisement; the contracting authority uses a negotiated approach to invite suppliers to submit ideas to develop innovative works, supplies or services, aimed at meeting a need for which there is no suitable existing 'product' on the market. The contracting authority is allowed to award partnerships to more than one supplier. The creation of innovation partnerships enables a public authority to enter into a structured partnership with a supplier with the objective of developing an innovative product, service or works, with the subsequent purchase of the outcome.

In deciding upon the appropriate procedure for a works contract, certain questions need to be addressed – see Fig. 6.1. The figure shows the flowcharts for the different approaches, including the time restrictions.

Transparency

Transparency is ensured by the publication of notices on public contracts in the EU's Official Journal and TED database as well as at national level. All publications must contain identical information so as not to favour any bidder, such as:

- Deadlines for bids
- Language(s) of bid
- Award criteria and their relative weighting
- Certificates/documents to accompany bids to allow the evaluation of a candidate's suitability to perform a contract.

Tender pre-qualification and evaluation

The aim of a PQQ is to check that the bidding company is (for the contract term):

- Technically competent
- Financially sound
- Legally compliant.

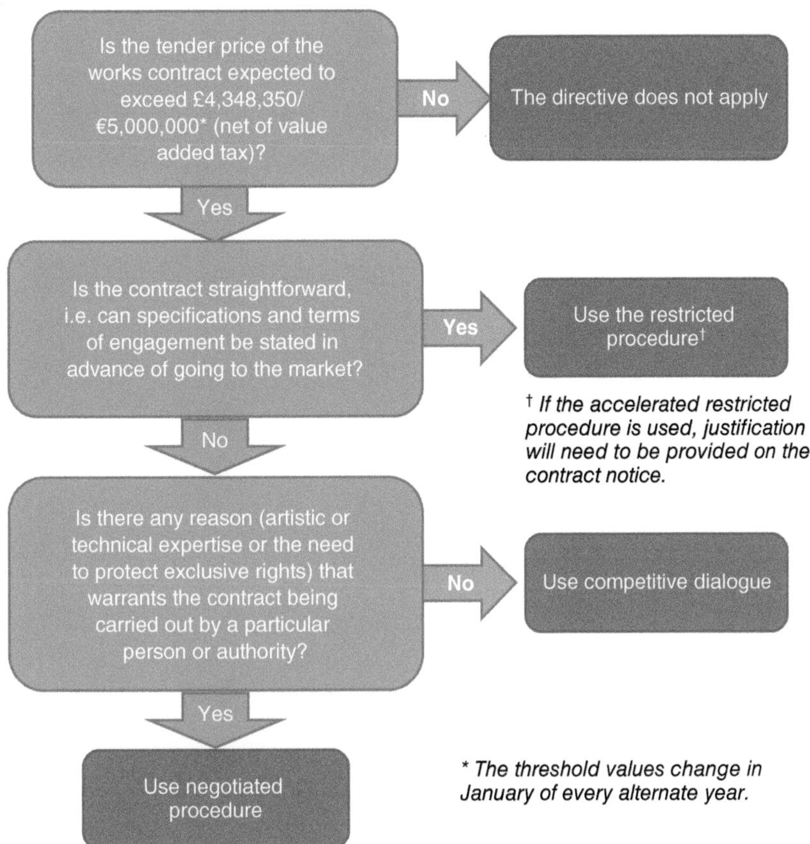

Figure 6.1 The flowcharts for the restricted and competitive dialogue procedures.

Figure 6.2 shows the evaluation criteria in more detail, split between commercial and technical. Contracting authorities are required to publish an award notification letter providing some information to unsuccessful tenderers about the reasons for their decision. After the contracting authority has informed tenderers of its decision, there must follow a 10-day standstill period during which the contract cannot be awarded, to ensure there is no appeal.

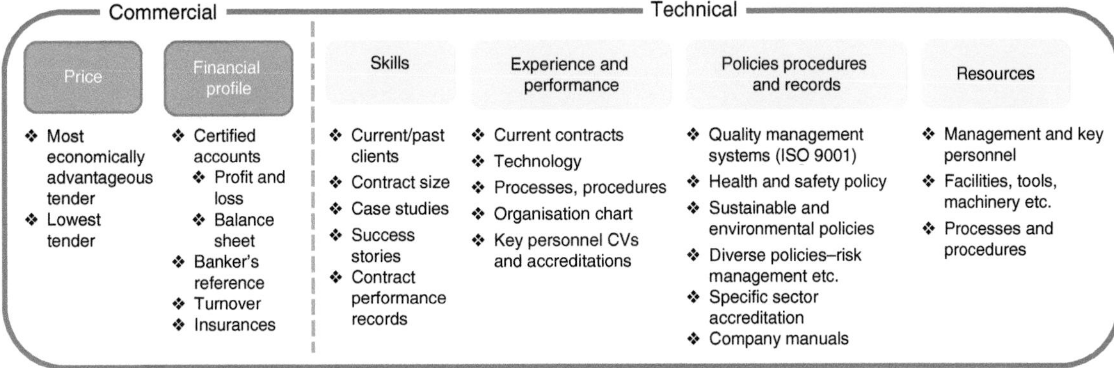

Figure 6.2 The criteria considered in a PQQ.

7 Procurement, selection, contractual arrangements and legal issues

7.1 Methods of procurement

Procurement is essentially a series of considered risks – each method has individual strengths and weaknesses. Procurement, being a series of risks, has different methods that transfer varying levels of risk onto the client or contractor. The estimating/bid team face a wide array of different procurement methods. Contractors must bid to win work, not just by being the lowest bidder on price. Some clients select on lowest price, health and safety record, environmental sustainability, quality assurance, innovation and ethical responsibility.

New methods of procurement are constantly evolving, with new forms of contract embodying different approaches to the management of risk. Figure 7.1 shows the different methods of procurement.

The five main methods of procurement are:

- Traditional/conventional, where design is separated from construction
- Design and build (D&B)
- Management procurement, such as construction management and management contracting where design and production can proceed in parallel
- Integrated, sometimes known as collaborative procurement, partnering, alliancing, where the focus is upon collaboration and working together
- Concession agreements, such as build/operate/transfer (BOT), public–private partnerships, private finance initiatives, where a team is formed for project delivery over a concession period award by the sponsor.

BS 8534 sets out the principles of procurement. Consideration should be given to:

- Issues of bribery and anti-competitive practice
- Dispute resolution
- Methods of identification and management of risk
- Issues of payment and financial management
- Corporate social responsibility
- Health and safety

All procurement methods are seeking approaches to balance the cost, time, quality and risk on a project.

New Code of Estimating Practice, First Edition. The Chartered Institute of Building.

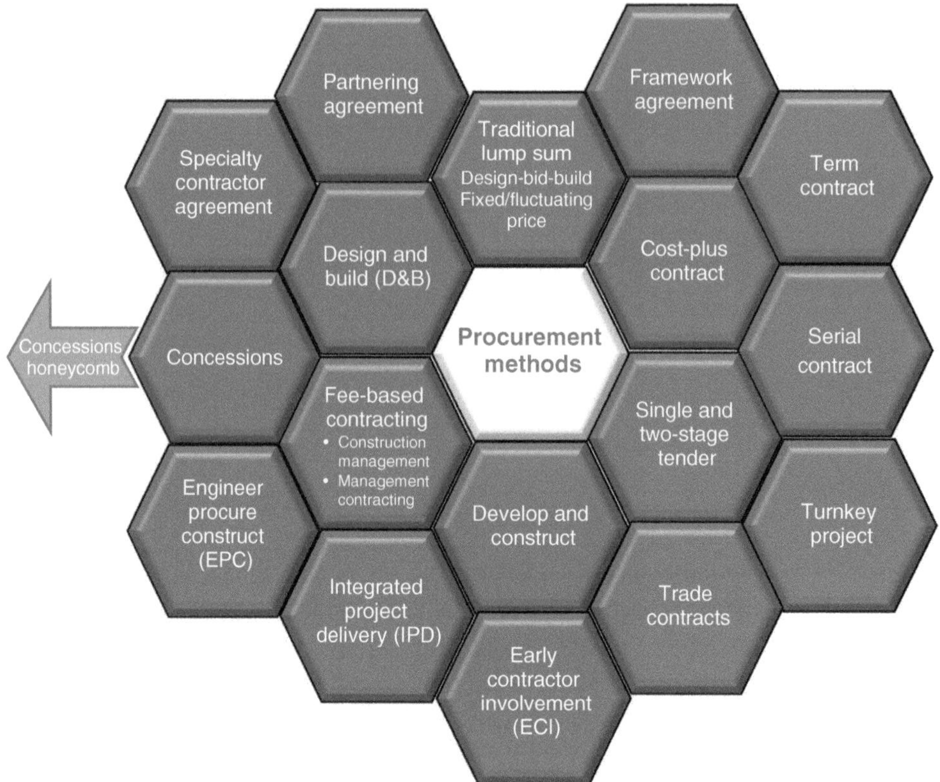

Figure 7.1 The procurement honeycomb.

- Environmental sustainability
- Intellectual property
- Conflict of interest
- Attitude to risk allocation.

In addition, consideration must be given to:

- Public procurement rules
- Tax and VAT
- Planning requirements.

The Code of Estimating Practice (CoEP) focuses principally upon conventional lump-sum methods of procurement, using a contract such as the Joint Contracts Tribunal (JCT) Standard Building Contract. D&B is growing in popularity as a procurement route; the CoEP has therefore included more detail on this method of procurement, with the implication for estimating. Many other procurement approaches are widely used.

Figure 7.2 shows the main procurement methods.

Traditional/conventional (design, bid and build) lump-sum projects

The separation of design and construction using a lump-sum contract has its weaknesses as all other methods of procurement do. However, the construction industry has used the traditional process for so long that it has become the most understood. The simplicity involved in understanding traditional is its greatest strength – the

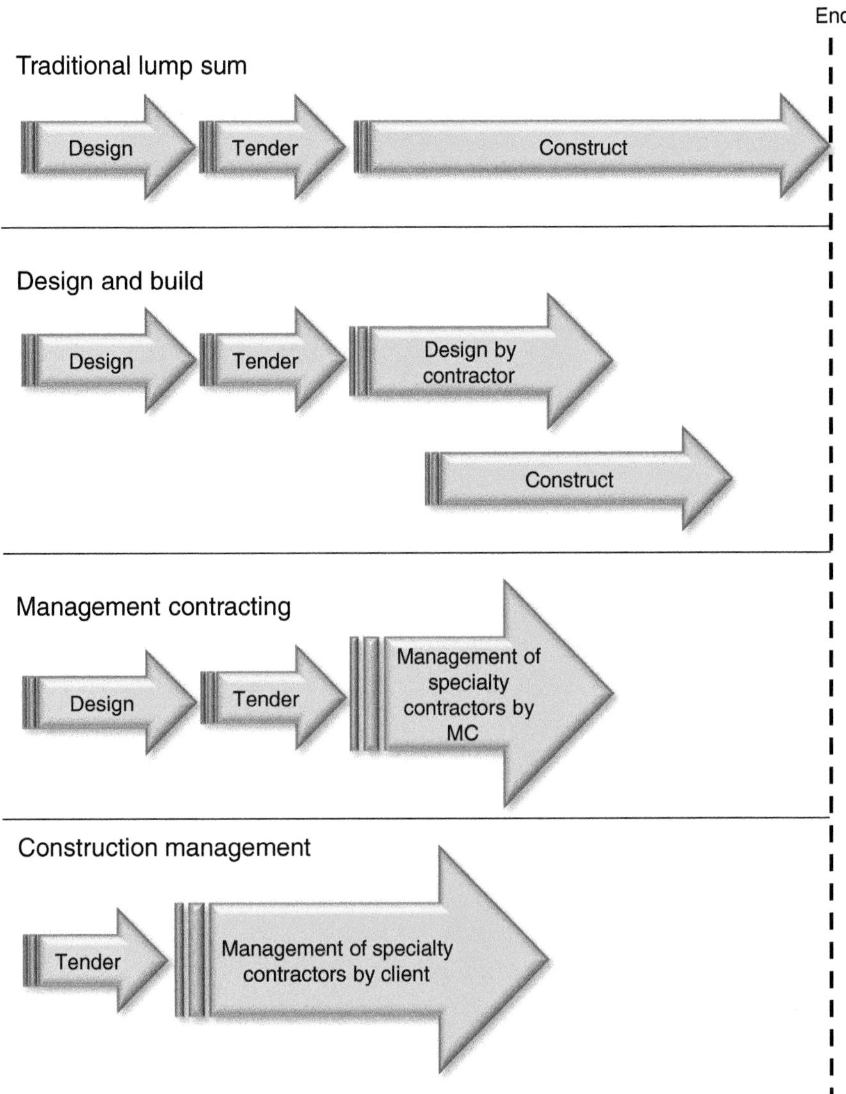

Figure 7.2 Procurement routes. *Source:* RIBA (2013).

BS 8534 2011 Construction procurement policies, strategy and procedures - code of practice - provides recommendations for procurement policies, strategy and procedures and managing performance and delivery for tendering.

BS 8903 Principles and framework for procuring sustainably.

BS EN ISO 9001 Quality management systems - requirements

designer is responsible for design and the contractor for execution, so responsibility for co-ordination of specialty contract packages lies firmly with the contractor.

While complications will inevitably arise, as with any procurement system, the traditional method sees each party knowing where they stand, and who has responsibility for what.

Separating the design and construction processes means disputes are common. Traditional construction involves projects with drawings and specification, drawings and bill of quantities with quantities/approximate quantities, and cost plans used as the basis for negotiating the tender price.

Full documents are required for the tendering process, whereas, in reality, the design is often incomplete. Traditional/conventional can also include the contractor's design, which requires special clauses in the contract. It is appropriate when the contractor is required to design specific parts of the works.

This type of contract requires the contractor to have professional indemnity insurance for the design portion of the work.

BS ISO 10845-1 2010 Construction procurement – processes, methods and procedures – established processes, methods and procedures for setting up a procurement system.

Measurement contracts are where the contract sum is not finalised until completion. This is appropriate when the scope and measurement cannot be competed until the work is undertaken, such as in a complex refurbishment project. The contract is initially likely to be based upon drawings and approximate quantities with re-measurement upon completion. The unit price rates will be based upon pro rata rates in the approximate quantities bill.

Cost reimbursement contracts are sometimes called cost-plus or prime-cost contracts. The contractor is reimbursed the actual cost of labour, materials and plant with an agreed allowance for overheads and profit. This approach is high risk for the client and low risk for the contractor.

Design and build (D&B)

D&B arrangements are popular with clients, as the risk primarily lies with the contractor, and the process is relatively easy to understand – the project is specified to be designed (at least in part) and built by the same contractor, which, in theory, allows for greater communication. Other parts of the design phase may be carried out by consultants hired by the client.

BS ISO 10845-2 2011 Construction procurement. Formatting and compilation of procurement documentation - provides general principles.

D&B has grown in importance with single point responsibility, the contractor taking responsibility for the design and construction of the project on a fixed price or guaranteed maximum price that gives the client some flexibility on design changes during site production. Consideration must be given to the appropriateness of using D&B, having regard to the novelty and complexity of the project. D&B can be a turnkey arrangement with the contractor taking responsibility to deliver a fully equipped and operational project.

The client's requirements should be clearly set out, upon which the contractor's D&B tender proposal is based. Adequate time is needed for bid preparation, because of the design component. With D&B contracts, the bidder must be very careful about terms that pass the burden of the unknown onto the contractor.

D&B selection

Selecting the successful D&B contractor is often on a two-stage approach. Submitting a D&B bid is expensive because of design fees and the work involved for contractors. The NJCC* Code of Procedure for Selective Tendering for Design and Build recommended that for single-stage tendering, no more than three firms should be invited to bid, and for two-stage tenders, no more than five D&B firms at stage one. In some cases, the client will offer to reimburse a predetermined fixed cost for the design component of unsuccessful bids, which is unlikely to cover the full cost that has been incurred by the contractor; it is payable after a successful tender has been accepted.

The bid team

Estimating D&B projects involves a bid team, which can include a design manager. D&B can be either one- or two-stage D&B. The design team appointed by the client for stage one can be novated to the contractor for the stage two bid, or the contractor can appoint their own design team. The design period and lead time can be shortened by overlapping design with construction where the tender process permits it.

* National Joint Council for the Building Industry, Code of procedure for Selective Tendering for Design and Build, NJCC, 1999, has been replaced by the Joint Contracts Tribunal, Practice Note 2012 Tendering.

The benefits are perceived to be more cost-effective projects, as the contractor can apply expertise to use the most buildable and cost-effective design solutions, materials and methods to achieve the client's requirements by managing the design process and taking account of critical production issues. The client benefits by reduced risk exposure by passing some of the design risk to the contractor.

D&B brings significant challenges to the pre-construction process. The drive to find economical design solutions can involve the contractor in taking risks to provide the requirement of 'fitness for purpose' for the client. Design risk for some projects can prove to be very considerable. Compressed and overlapping design and construction programmes can lead to significant management problems with design delays primarily at the contractor's risk.

Contractors must bring additional resources and procedures into the pre-construction process to meet these challenges. Risk management is of particular importance. For major D&B projects it is wise to employ an independent risk manager with a background in design risks. Design can be carried out by an internal design department, external design practices or by specialty contractors. The best solution will depend on the company and the project. Only major companies with a regular flow of D&B work can afford to maintain a design department, so sub-contracting of design and drawing production is a very common solution.

The design team for a D&B project will usually be assembled specifically to work during the tender period on a selected project. The make-up of the team will vary depending on the nature of the project but typically includes the estimator, planning and scheduling engineer, production/site manager designate (to cover the buildability practicalities), design and technical manager (to co-ordinate the design), purchasing and supply representative, together with the design specialists (architect, structural engineer, mechanical and electrical and plumbing specialist).

One of the contractor's representatives will lead the team, often referred to as the 'bid manager', and for the architect to assume the role of 'lead consultant'. This assumes that the contractor is at liberty to choose his own consultants: this is not always the case as they can be novated.

During design team meetings, the estimator's role should be pro-active. The estimator should challenge the designers to produce cost-effective solutions and not simply to take on a clerical role. Where the estimator sees excessive cost, the team should be made aware of it, as it is generally the most cost-effective scheme that wins, not the most extravagant.

The estimator should also make the team aware of the importance of maintaining the key dates on the tender preparation programme for the provision of design information. This ensures that specialty-contractor and supplier enquiries can be issued on time and that bill of quantities preparation can be managed within agreed periods. Value engineering should be an area where all team members are encouraged to participate. Each idea should be given a unique identification number and recorded on a list. The estimator can then approximately cost each one, discarding those which are not effective, and proceed to produce detailed costs for the remainder. The estimator can then present the list at the tender settlement, and decisions can be made regarding inclusion, or otherwise, in the eventual tender.

Due to the wide range of specialists attending and the complexities of the project, design team meetings need to be structured and conducted to a timetable.

Defining the consultant's role and agreeing fees

Consultants will be well versed in agreeing briefs and fees

Defining the details and responsibilities

The prospective consultant must be given a clear and precise brief regarding what services are to be provided and what responsibilities are to be carried. Fringe matters, such as the number of copies of each drawing to be issued, the consultant's involvement with on-going site inspections, frequency of site meetings and who is to attend, will all affect the fee and should be detailed in the brief.

Dates for issue of design information, which will dictate the levels of resources that the consultant will need to consider, together with details of warrantee requirements, and levels of professional indemnity insurance, also need to be defined. The consultant should be clear as to their role and associated responsibilities.

If the consultant is to provide services during the tender period on a 'no job–no fee' basis, this should also be clearly stated, and the value and recovery of those costs, should the tender be successful, incorporated into the post-tender fee structure.

It is desirable to have a 'design responsibility matrix' listing all elements that have design input, with a clear statement of who has what level of responsibility for them under one of the following categories:

- Total responsibility
- Primary responsibility, with input from those with contributory responsibility
- Contributory responsibility, led by those with primary responsibility.

Management procurement/fee-based contracting†

Construction management and management contracting are fee-based/agency methods of construction procurement, with many derivatives, such as construction management at risk with a guaranteed maximum price.

In management contracting, the management contractor undertakes to manage carrying out of the work through specialty contractors who are contractually accountable to the management contractor. Projects can be awarded on the basis of target cost and cost plus reimbursement.

Construction management

This is mainly used on large and/or very complex construction works. It involves having a construction manager as a point of contact, who will typically be the head of a design team and who co-ordinates the project in terms of the various construction operations on-site. Construction Management is generally considered to be the least adversarial form of procurement and is often used when design needs to run in tandem with construction.

Develop and construct

A variant of D&B, known as Develop and Construct, where the client has procured a concept design before passing the design responsibility to the contractor. The contractor develops the design to detail stage which is submitted to the client. This gives the client greater control over the design but remains within the D&B concept.

† Fee based contracts are becoming more common internationally, with China using the term Agent Construction System (ACS).

Management contracting

Management contracting works by having a contractor managing a series of work package contractors. Advantages include early involvement in the project, and the management contractor can also appoint trusted specialty contractors they have worked with previously rather than risk an unknown factor. Disadvantages include the lack of a single point of responsibility for both design and construction phases; this opens the possibility for disputes to arise.

7.2 Two-stage tendering

In the two-stage tender process, the contractor becomes involved in the planning of the project at an earlier stage; the tenders are based on minimal information at the first stage, such as a schedule of rates for key measured items. In the second stage, the client's team will develop the precise specification in conjunction with the preferred tenderer. This method is favoured for complex projects, where the contractor may have significant design input. Contractually, in the first stage, the contractor submits the bid with an agreement to enter into second-stage negotiations and a pre-stage services agreement, subject to satisfactory negotiations on the priced bill of quantities for the second stage. Sometimes two contractors will be selected to work on the second stage, in the event that one of the contractors withdraws, Contractors prefer two-stage tendering because it reduces abortive bidding and provides time to identify and allocate risks.

The advantages of two-stage tendering are:

- The saving in the cost of abortive bidding
- The client team can receive the benefit of buildability/constructability input to the design process and value engineering the project
- The ability to select design solutions in collaboration with the supply chain.

Two-stage tendering may use a cost plan to be priced by the contractor in preference to a fully measured bill of quantities.

At the first stage, the appointment needs to be made based on a tailored agreement that sets out all the tender items to be applied to the construction contract. However, the contract does not place an obligation on the contractor to proceed to the 'main' construction contract. If they don't, settlement is made based on a pre-construction fee which may be based on collaboration with the design and/or consultant teams and the procurement of prices (on an open-book basis) for work-packages from sub-contractors and suppliers. Appointment at the first stage may require:

- A pre-construction and construction programme
- Method statements
- Detailed preliminaries including staff costs
- Agreed overheads and profit
- A schedule of rates to be applied to the second-stage tender
- Agreed fees for design and other pre-construction services.
- CVs for proposed site and head office staff
- Tendering of any packages that can be broken out and defined
- Agreed contract conditions to be applied to the second-stage construction contract.

Design risk may be transferred from the client to the contractor under this form of tendering. But, on the other hand, the client loses some control over the design. Two-stage tendering does present different risks. The cost of two-stage tendering may be higher, but the likelihood of expensive variations and/or claims is lower. Proposals must be robust, with the second-stage process being clearly articulated, any ambiguity will slow down the bid process. Relationships are improved over the longer period of collaboration, which can improve performance. It can reduce competitiveness as others will be less keen to enter at the second stage if another contractor is already in place at the first stage.

7.3 Framework agreements

Framework agreements, partnering and alliance contracting are all approaches to procurement.

A framework agreement is a general term for agreements with providers that set out terms and conditions under which specific purchases (call-offs or projects) can be made throughout the term of the agreement. In most cases, a framework agreement is not a contract but the procurement for one or more contracts on an ongoing basis.

Partnering is a management approach used by two or more organisations to achieve specific objectives by maximising the effectiveness of each participant's resources. The approach requires mutual trust and an open relationship between parties and results in a combined drive to achieve improvement and best practice. The principles of partnering include a decision-making process, mutual objectives and overall improvement in performance.

7.4 Concession contracts

A concession gives a concessionaire the long-term right to use all utility assets conferred on them, including responsibility for and some investment. Figure 7.3 shows the different types of concession contracts, with the major types on the right shown highlighted.

Asset ownership remains with the authority, and the authority is typically responsible for replacement of larger assets. Assets revert to the authority at the end of the concession period, including assets purchased by the concessionaire.

The concessionaire typically obtains most of its revenues directly from the consumer and so has a direct relationship with the consumer. It covers an entire infrastructure system (so may include the taking over of existing assets as well as building and operating new assets). The concessionaire will pay a concession fee to the authority which will usually be ring-fenced and put towards asset replacement and expansion.

The EU Directive 2014/23 on the award of concession contracts (the Concessions Directive) was introduced in 2014. It set out the rules for procurement by public sector contracting authorities and by contracting entities in the utilities sector by means of a concession and applies to concessions the value of which is equal to or greater than €5,186,000. This value shall be 'the total turnover of the concessionaire generated over the duration of the contract, net of VAT, as estimated by the contracting authority or the contracting entity, in consideration for the works and services being the object of the concession as well as for the supplies incidental to such works and services'. Member states were given 2years to transpose it to national law.

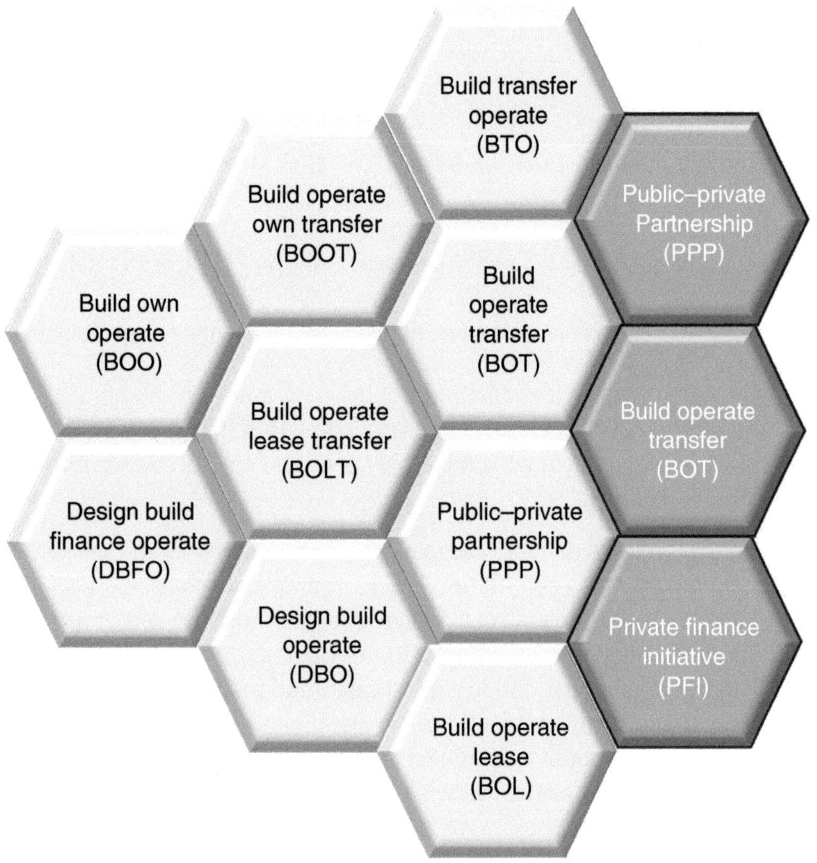

Figure 7.3 The different types of concession contracts.

Risk in the different types of procurement

Figure 7.4 shows the apportionment of risk for the different types of procurement. Risk management is discussed at greater length later in the Principles section.

Public–private partnership/ private finance initiative projects (PPP/PFI)

Governments are seeking to maintain infrastructure development and ways of financing through public/private partnerships and different procurement methods such as private finance initiative (PFI), public private partnership (PPP), build operate and

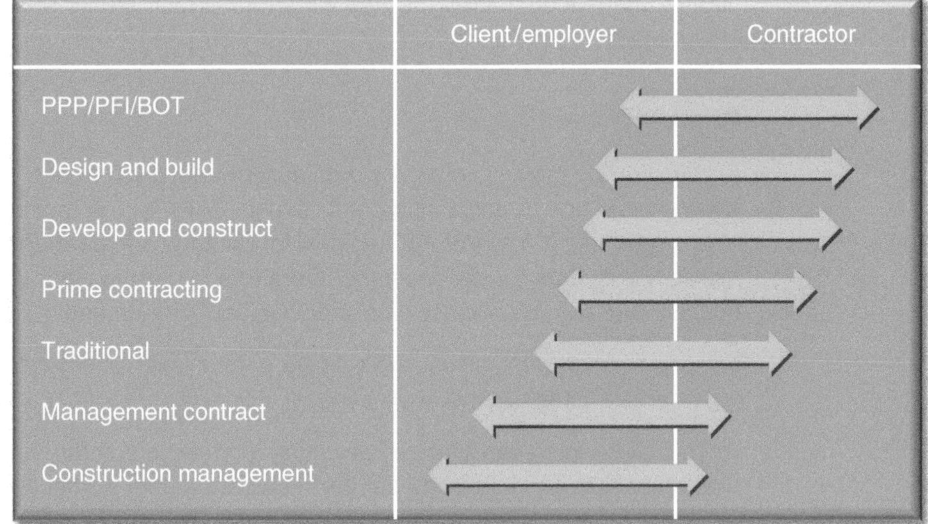

Figure 7.4 The apportionment of risk for different procurement types.

- BOO, build-own-operate
- BOL, build-operate-lease
- BOLT, build-operate-lease-transfer
- BOOT, build-own-operate-transfer
- BRT, build-rent-transfer
- BTO, build-transfer-operate
- DBO, design-build-operate
- DBOM, design-build-operate-maintain
- DBM, design-build-maintain
- DBFO, design-build-finance-operate

Private responsibility
High contractor risk

Public responsibility
High public sector risk

Figure 7.5 The different types of concessions and their level of contractor/public sector risk.

transfer (BOT) and build operate lease transfer (BOLT). There are a number of derivatives of BOT, each one along a continuum of public and private responsibility/risks – see Fig. 7.5.

Such projects are very complex, expensive to bid for and involve the delivery organisation/consortia, sometimes called the Special Purpose Vehicle (SPV) formed especially for the project, to finance, design, construct and maintain a facility, such as a school, hospital, prison or road. In return for a revenue stream or a deferred interest in the project, the consortia will take responsibility for the design, delivery and operations and maintenance risk. The successful bidder may not necessarily be the cheapest but the most economically advantageous tender. Because of the high bidding costs, bidders are sometimes reimbursed their bidding costs.

BOT

BOT is a project delivery method where the private sector entity (project company/SPV) is granted a concession by a government (host government/project sponsor) to design, finance, construct and operate a facility for a specified period (concession period) usually between 25 and 40 years, and then the facility is transferred free of charge to the host government.

7.5 Engineer Procure Construct (EPC)

Engineer Procure Construct (EPC) projects are a type of D&B project, where the contractor is involved in the design and construction of the facility as well as the design and procurement of the equipment and machinery. The equipment and machinery constitute a substantial portion of the total project cost. Petrochemical plants, pharmaceutical plants and wafer-fabrication plants are typical examples of EPC projects.

7.6 Prime contracting

A definition of prime contracting adopted by the UK Defence Estates is: 'A Prime Contractor is one having single point responsibility for the management and delivery of a project using a system of incentivisation and collaboration working to integrate the activities of its supply chain members to achieve a project that is on time, within budget, and is in accordance with the specified outputs and is fit for purpose.'

Most importantly, the focus is on single point responsibility and the ability to bring together and integrate the supply chain.

Factors taken into account when selecting a prime contractor include project management capability, financial standing, technical competence, supply chain arrangements, high ethical principles and a willingness to share risk. Collaborative working and relationships are important principles.

7.7 Early contractor involvement (ECI)

Early engagement of a contractor has become more important. Early contractor involvement (ECI) is a form of negotiated tender with emphasis placed on the contractor acting as the lead designer from the outset of the project. This is used mainly in infrastructure projects where the early appointment of a contractor can significantly influence the emerging design.

ECI involves the creation of a designer/contractor team led by the contractor. Contractor selection is similar to two-stage tendering as it is not on a cost basis but on the quality of the team and bid. It is beyond the scope of the CoEP to address ECI in detail, but its uses are widening.

7.8 Integrated Project Delivery (IPD)

Integrated Project Delivery (IPD) started in the United States of America and is now being used internationally.

The definition is: '*Integrated Project Delivery (IPD) is a project delivery approach that integrates people, systems, business structures and practices into a process that collaboratively harnesses the talents and insights of all participants to reduce waste and optimize efficiency through all phases of design, fabrication and construction. Integrated Project Delivery principles can be applied to a variety of contractual arrangements. Integrated Project Delivery teams will usually include members well beyond the basic triad of owner, designer and contractor. An integrated project includes tight collaboration between the owner, architect/engineers, and builders ultimately responsible for construction of the project, from early design through project handover.*' (*Source:* American Institute of Architects.)

The key to successful Integrated Project Delivery is assembling a team that is committed to collaborative processes and is capable of working together effectively. IPD means that the estimator will work alongside the design team to deliver a fully integrated approach throughout the design and production process.

7.9 Selection processes

Negotiation

A contractor may be approached to negotiate a tender price without the need to introduce competitive tenders. In such cases, a contractor may be selected on the basis of past performance, recommendation, familiarity with the work or most commonly because of a close working or business relationship with the client or consultants. This is more prevalent within the private sector. Negotiation allows early contractor selection, especially where the design would benefit from the constructor's input. This can reduce the overall project programme, increase buildability and tailor the costs to the client's budget. The counterargument is that the initial price may be higher and difficult to compare with competitive market rates. There are also procedures adopted by many (usually public) organisations to ensure that goods and services are procured by competitive means, particularly where finances are publicly accountable.

Open competitive tendering

This permits any applicant to join a lengthy tender list which can be, and often is, beyond a sensible level. This arrangement is an approved EU option. Open tendering is used where the lowest price is likely to be selected.

Approved lists and framework agreements

This form of selective tendering enables clients to choose tenderers for a project from a list of contractors who have been vetted and pre-qualified for various categories of work at an earlier stage. Contractors are asked to apply for categories, which are defined by contract value and nature of work. In a framework agreement there may be some guarantee of the number of enquiries to be sent to a contractor in a given period of time (can be 1–5years). It is important that clients and their representatives monitor and regularly update their lists of contractors to:

- Exclude companies whose performance has been unsatisfactory
- Introduce suitable new companies that can demonstrate the required qualifications and abilities
- Compile the lists in a form appropriate to the class of project
- Include firms with the financial capacity and stability to do the work.

One-off project (ad hoc) lists

Clients, or their consultants, frequently create an initial list of suitable contractors solely for a particular project. There will often be a pool of contractors from which to choose that can be assembled in three ways:

1. By including contractors who write with an early expression of interest in a scheme
2. By using an advertisement to invite applications or
3. Clients or consultants may draw on their experience or interrogate their database.

Contractors included on an initial list are usually asked to provide information about their financial and technical performance, particularly in relation to the type of work under consideration. More elaborate pre-qualification practices include completion of questionnaires and making presentations to the clients and their consultants. An assessment of a contractor's competence in compliance with health and safety legislation is now a statutory requirement prior to the award of a contract and so may be dealt with at the pre-qualification stage. The EU procurement rules are more prescriptive as to what information should be provided by contractors about themselves.

Evaluation strategy

Contracts are awarded on the basis of: The most economically advantageous tender (MEAT); or the lowest price.

The criteria on which the tenders will be evaluated must be detailed in the tender document. Where the tender is to be based on MEAT, the relevant criteria should be included, such as:

- Quality
- Price

- Technical merit
- Aesthetic and functional characteristics
- Environmental characteristics
- Running costs
- Cost-effectiveness
- After-sales service and technical assistance
- Delivery date and delivery period or period of completion.

7.10 Integrated design and construction

Introduced by the CIOB in 2015, the integrated design and construction single responsibility (IDCsr) integrates within a single team all the key participants involved in the process of designing and constructing a project. IDCsr is a totally integrated yet competitive form of procurement, design and project delivery approach. Its over-riding principle is that of purchasing a customised product at a fixed price, rather than the traditional commissioning of a contract service. It involves integrating design and construction through a single legal entity, the IDCsr constructor, as in any other product manufacturing sector.

It is different to D&B as IDCsr is divided into three stages:

- *Stage 1*: Inception
- *Stage 2*: Selection
- *Stage 3*: Delivery.

The integrated design and construction contractor (IDCsr) is involved from the outset, with helping the client to develop the business case, develop the project definition and client requirements and preparing all documentation. The IDCsr accepts total, single responsibility for the entire design and construction process as well as the finished product. All the mechanisms within traditional construction contract forms are designed to apportion authority, responsibility and blame; but with the IDCsr Sale Agreement, these approaches are inappropriate and redundant. The IDCsr's Sale Agreement has been prepared as a standard product sale agreement.

7.11 E-procurement

All procurement documentation must be available via the internet from the date of publication of the contract notice (with some exceptions). EU Member states have some freedom in making policy decisions concerning electronic requirements on public contracts. The United Kingdom has chosen:

- To postpone mandatory use of electronic communication and the requirement on contracting authorities to use 'e-Certis' until 18 October 2018.
- Not to set centrally the level of security in e-communications but instead to leave this to each contracting authority within a centrally set 'framework' as set out within the Regulations, including the levels of risk requiring the use of advanced electronic signatures.
- Not to mandate the use of electronic catalogues or the use of 'building information electronic modelling' for works contracts.

7.12 E-auctions

Electronic auctions may be used and should be based on:

Either

solely on prices when the contract is awarded to the lowest price

Or

on prices and/or on the new values of the features of the tenders indicated in the specification when the contract is awarded to the most economically advantageous tender.

The contract notice for an e-auction should include:

- The features, the values for which will be the subject of electronic auction, provided that such features are quantifiable and can be expressed in figures or percentages;
- Any limits on the values which may be submitted, as they result from the specifications relating to the subject of the contract;
- The information which will be made available to tenderers in the course of the electronic auction and, where appropriate, when it will be made available to them;
- The relevant information concerning the electronic auction process;
- The conditions under which the tenderers will be able to bid and, in particular, the minimum differences which will, where appropriate, be required when bidding;
- The relevant information concerning the electronic equipment used and the arrangements and technical specifications for connection.

Before proceeding with an electronic auction, contracting authorities shall make a full initial evaluation of the tenders in accordance with the award criterion/criteria set and with the weighting fixed for them. All tenderers who have submitted admissible tenders shall be invited simultaneously by electronic means to submit new prices and/or new values; the invitation shall contain all relevant information concerning individual connection to the electronic equipment being used and shall state the date and time of the start of the electronic auction. The electronic auction may take place in a number of successive phases. The electronic auction may not start sooner than two working days after the date on which invitations are sent out.

Article 35 of the Directive states that 'contracts with intellectual performances shall not be the object of electronic auction'. Services like engineering, consultancy, architecture or design are covered by the collective description of intellectual performances or services.

7.13 Abnormally low tenders

If tenders for a contract appear to be abnormally low in relation to the goods, works or services, the contracting authority shall, before it may reject those tenders, request in writing details of the constituent elements of the tender which it considers relevant.

8 Preliminaries

Pricing the preliminaries will depend upon decisions made in terms of duration, timing and methods, contained in the **Method statement**. Other decisions such as major plant requirements, gang sizes and choice of speciality contractors also feed into the pricing of preliminaries. **Temporary works** are a significant part of the bid cost and are dealt with in a separate section. **Health and Safety**, which impinges on a number of honeycombs is also dealt with in a separate section.

The honeycomb in Fig. 8.1 shows the items involved in pricing the preliminaries. Some of these are described in more detail, and others are covered elsewhere. Four preliminaries' items make up around 80% of total costs: staffing, mechanical plant, access/scaffolding and site accommodation – see Fig. 8.2.

8.1 Site establishment

Site establishment is an important part of the project process and leads to an efficient site with good productivity. Figure 8.3 shows the site establishment honeycomb. Estimating for this part of the project can be crucial if the optimum facilities are to be provided. The main considerations are:

- Accommodation
- Security
- Safety
- Site assessment
- Temporary utilities.

A lot of information forms the basis of the estimating process for site establishment, including:

- Bill of quantities
- Method statement
- Site access plan
- Form of contract
- The drawings and specifications and requirements

New Code of Estimating Practice, First Edition. The Chartered Institute of Building.

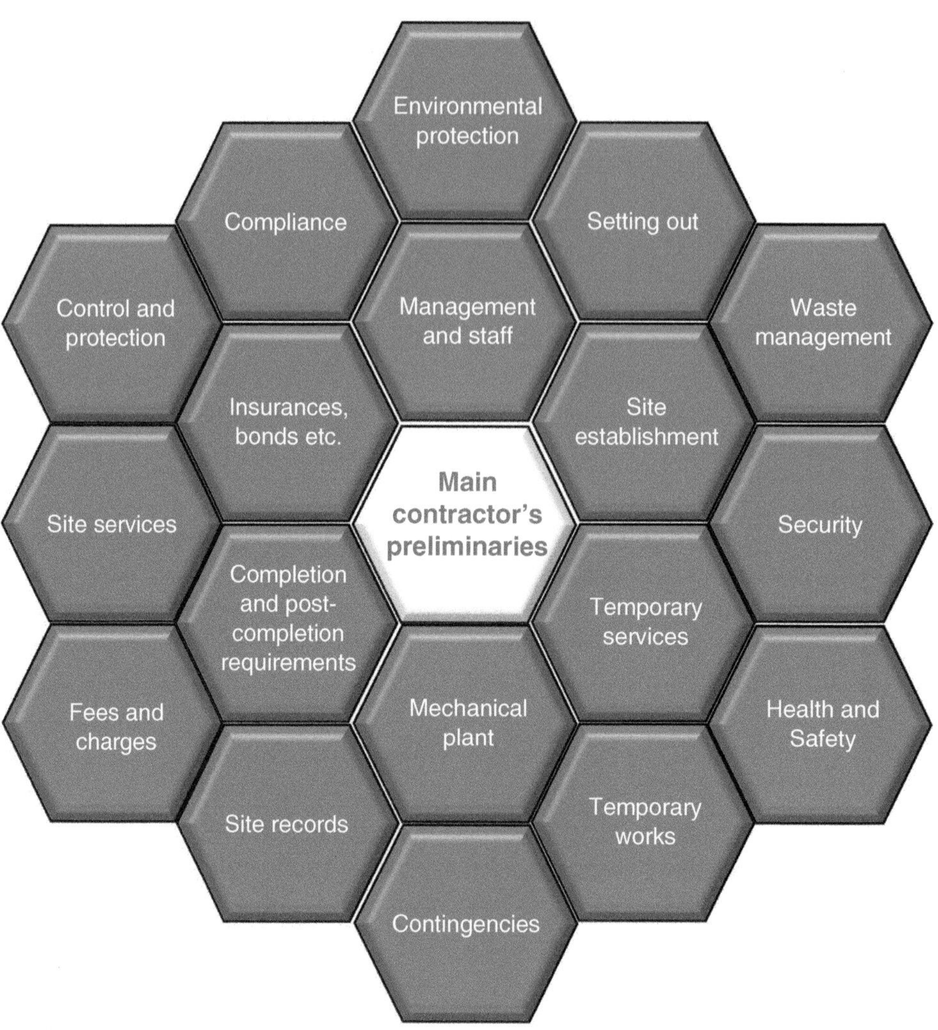

Figure 8.1 The preliminaries honeycomb.

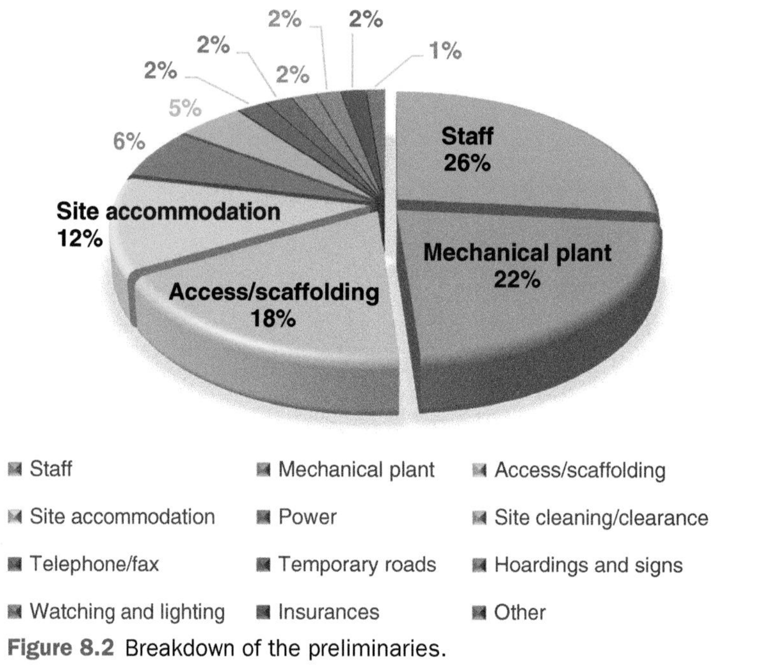

Figure 8.2 Breakdown of the preliminaries.

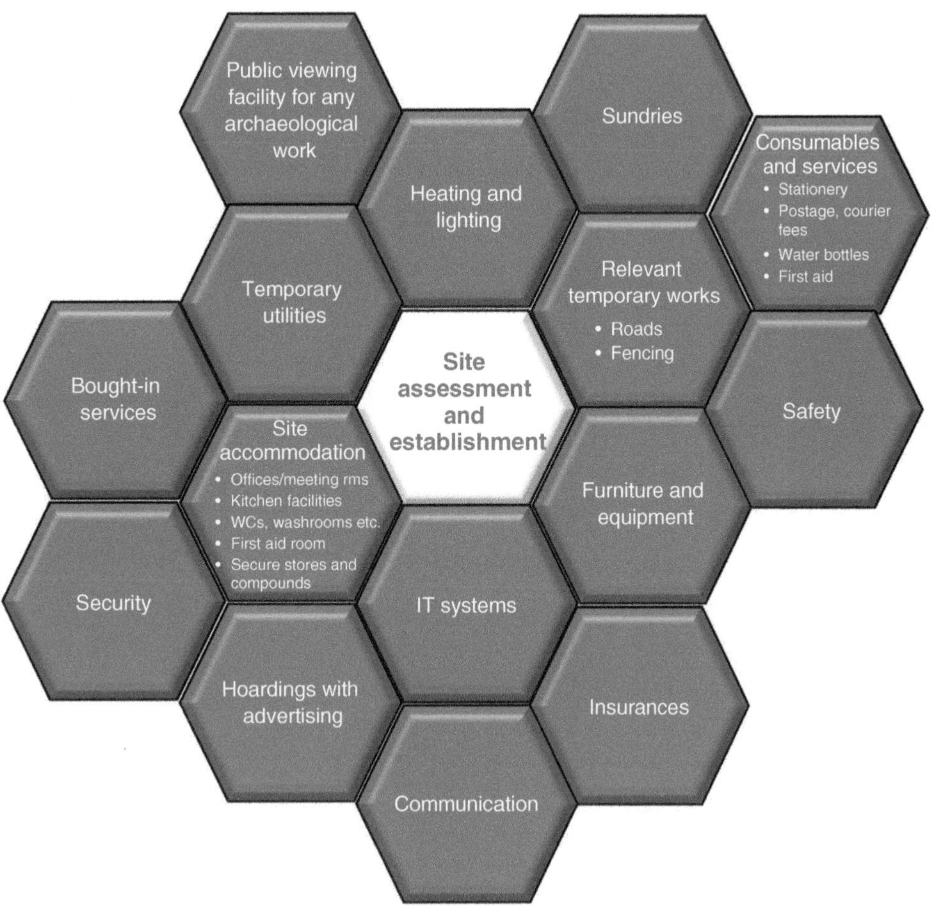

Figure 8.3 The site establishment honeycomb.

- Site investigation report
- Resource availability
- Safety/risk assessments
- Specialty contractor availability
- Site waste management plan
- Preliminaries and individual site conditions and location
- Temporary works design
- Contract programme
- Prime costs and provisional sums
- Construction phase plan (large projects)
- Specialty contractors' needs, for example office space, parking, power requirements and water.

The following sections provide details of each of the hexagons in the site establishment honeycomb:

1. Site accommodation
2. Heating and lighting

3. Hoardings with advertising
4. Security
5. Communication infrastructure
6. Bought-in services
7. Consumables and services
8. Public viewing
9. Sundries.

Site accommodation

The CDM regulations (2015) outline the minimum welfare requirements for site personnel (HSE 2015b). They are:

- Sanitary conveniences
- Washing facilities
- Drinking water
- Changing rooms and lockers
- Rest facilities.

Sanitary conveniences The conveniences must be provided or made available in readily accessible places. They must be adequately ventilated and lit, and separate rooms must be provided for men and women.

Washing facilities Suitable and sufficient washing facilities, including showers if required by the nature of the work or for health reasons, must, so far as is reasonably practicable, be provided or made available at readily accessible places. They should be in the vicinity of every sanitary convenience and should include a supply of clean hot and cold, or warm, water, soap or other suitable means of cleaning and towels or other suitable means of drying. Rooms containing washing facilities must be sufficiently ventilated and lit.

Drinking water An adequate supply of wholesome drinking water must be provided or made available at readily accessible and suitable places. Where necessary, for reasons of health or safety, every supply of drinking water must be conspicuously marked by an appropriate sign. Where a supply of drinking water is provided, a sufficient number of suitable cups or other drinking vessels must also be provided, unless the supply of drinking water is in a jet from which persons can drink easily.

Changing rooms and lockers Suitable and sufficient changing rooms must be provided or made available at readily accessible places if a worker has to wear special clothing for the purposes of construction work and cannot, for reasons of health or propriety, be expected to change elsewhere. Where necessary for reasons of propriety, there must be separate changing rooms for, or separate use of rooms by, men and women. Changing rooms must be provided with seating and have drying facilities. Suitable and sufficient facilities must, where necessary, be provided or made available at readily accessible places to enable persons to lock away any special clothing which is not taken home; their own clothing which is not worn during working hours and their personal effects.

Rest facilities

The rest facilities should be maintained at an appropriate temperature and have:

- An adequate number of tables and chairs
- A means of heating food (e.g. a gas or electrical heating ring or microwave)
- A means of boiling water.

Accommodation on the site is usually temporary, unless there is available space in a permanent facility. The accommodation may include offices, meeting rooms and first-aid room. Elsewhere on-site, facilities are needed for storage in the form of either compounds or metal storage units. Standard site offices are available in four sizes:

Length: 3.6 m, 4.8 m, 6.0 m and 7.3 m; Width: 2.4 m; Height: 3 m approx.

A number of considerations need to be taken into account in estimating temporary accommodation requirements, such as size, number, siting and accessibility. Restrictions may apply which need to be considered: Project-specific restrictions; width and height restrictions; traffic management requirements for delivery and removal; access/traffic restrictions off-site; vehicle restrictions on-site and ground conditions.

Secure stores and compounds

Security of on-site for materials, plant and equipment is important. Lockable or alarmed/patrolled facilities are necessary, appropriately weather proofed and access to them needs to be straightforward and clear of hazards.

Heating and lighting

Heating and lighting is required in temporary accommodation for the comfort of personnel and incurs:

- Energy and installation costs
- Connection to the temporary electricity supply/generator and maintenance.

Rental charges and any metering requirements should be allowed for where the heating or lighting is hired rather than purchased, and safety requirements for overhead cables that must be observed.

Heating - Site accommodation must be suitably heated, or cooled, where necessary.

Lighting - Lighting on the site is important for safety, productivity and quality. The need may be driven by either the working hours required or the local climate. The CDM Regulations (2015)(35) state that:

- Each construction site and approach and traffic route to that site must be provided with suitable and sufficient lighting, which must be, so far as is reasonably practicable, by natural light.
- The colour of any artificial lighting provided must not adversely affect or change the perception of any sign or signal provided for the purposes of health or safety.
- Suitable and sufficient secondary lighting must be provided in any place where there would be a risk to the health or safety of a person in the event of the failure of primary artificial lighting.

Floodlighting may be necessary for safety, security, productivity and quality. A number of factors need to be taken into account if floodlighting is used:

- Cost and duration
- Planning laws
- Positions where poles/towers can be placed
- The level of glare that is acceptable
- Ease of maintenance required

Emergency lighting should be provided on emergency routes and exits – see **Temporary works**.

Planning regulation – Class 8 (Schedule 3) Hoardings around Temporary Construction Sites
Relates to the display of advertisements on hoardings at temporary construction sites. The guidance states that any advertising: Will provide some environmental benefits; Must not be displayed more than 3 months before work starts; Is a maximum of 38 m² in area; Is not more than 4.6 m above ground level; Details must be sent to the local authority at least 14 days before display; May be illuminated in a reasonable manner.

Hoardings with advertising

Hoardings provide protection and security for a site. They are also an opportunity to advertise the contractor/client or other shareholders, thus offsetting the cost of the hoarding against the advertising revenue. There is a planning regulation for advertisements on hoardings which needs to be considered, either for compliance or to allow for the cost of a planning application (money and time).

Buildings which are being renovated or are undergoing major structural work and which have scaffolding or netting around them may be considered suitable as temporary sites for shroud advertisements or large 'wrap' advertisements covering the face, or part of the face, of the building. In all cases, express consent from the local planning authority will be required for these advertisements.

If the building to which the scaffolding is connected is included on the Statutory List of Buildings of Special Architectural or Historical Interest, listed building consent is likely to be required.

Security

The protection and security of a site may be a requirement of the client or will be based on a decision by the contractor. Protection may take the form of fencing/hoarding around the site perimeter. This will require planning permission and estimated based on:

- The purchase/hire of the hoardings
- Erection
- Maintenance
- Dismantling.

Site security involves labour costs and, if there is to be on-site security, consideration needs to be given to appropriate accommodation/shelter needs. Most large projects will have a security gate where materials are checked, and the movements of visitors, workers and staff are monitored. A digital clocking system will be used on very large projects, to check on movements on the site.

Communication infrastructure

Reliable voice, picture and data communications has become one of the most important items for any project. With CAD, BIM, e-mail, collaboration tools and transmission of digital data, the communications infrastructure is crucial to all projects.

A telephone connection, with associated broadband, is an important feature of construction sites, although mobile phones can be the choice of communication within and beyond the site. The broadband provider will provide the infrastructure to ensure continuous and reliable connectivity. Allowances must be made for:

- Initial site establishment of mobile and broadband facilities
- Monthly rental and usage charges, including allowances for excessive use
- Cloud facilities.

Bought-in services

These are services outsourced by the main contractor and include:

- Catering
- Equipment maintenance
- Document management, including management information systems and electronic data management systems (EDMS)
- Printing (purchasing), including reports and drawings
- Staff transport
- Off-site parking charges
- Meeting room facilities
- Photographic services.

Consumables and services

Consumables include:

- Stationery, postage and courier charges
- Computer, printer, fax and photocopier consumables
- Tea, coffee, water bottles and the like
- First-aid consumables.

Public viewing

Many sites, particularly those with high public interest, provide facilities for the public to view the site (e.g. Crossrail in London). These can be either platforms or 'windows' in the site perimeter hoarding. This is also the case for archaeological interest/finds.

Sundries

- Main contractor's signboards
- Safety and information notice boards
- Fire points

- Shelters
- Tool stores
- Crane signage
- Employer's composite signboards.

8.2 Insurances, bonds and so on

Insurance

Insurance is a way of managing risks on construction projects. The sums involved in a contractor's contract liability can be substantial – the higher the risk, the higher the premium. Insurance protects both the contractor and the client. The estimator needs to clarify, at a very early stage, which insurances are to be covered by the client and to be aware of any gaps between the client-owned insurance and that of the contractor.

It is common for companies to express their insurance costs as a percentage of turnover and to include the cost in the project overheads schedule when the full value of the project is known. Many different types of insurances are available to a contractor; the main ones are:

1. All-risks
2. Public liability
3. Employer's liability
4. Professional indemnity
5. Latent defects
6. Other insurances.

All-risks insurance All-risks insurance is a policy that provides coverage for both damage to a property and third-party injury or damage claims. This type of insurance covers any physical damage to the project works or materials on-site. It is usually taken out in joint names (contractor and client) so that, regardless of fault, the funds will be available. Each party retains the right to file a claim; all parties have a duty to inform the insurer of any injuries or damages that may lead to a claim.

Public liability insurance Public liability insurance is required to provide cover against personal injury or death, or loss or damage to property of third parties, such as members of the public or independent specialty contractors.

Employer's (main contractor's) liability insurance Employers' liability insurance covers the cost of compensating employees who are injured at, or become ill through, work. An employer is legally obliged to have employers' liability insurance and can be fined up to £2,500 for every day without appropriate insurance. Employers' liability insurance covers the cost of compensation and any associated legal fees.

Professional indemnity insurance

Professional indemnity insurance, also called professional liability insurance and errors and omissions insurance in the United States of America, helps to protect the service-providing individuals against the cost of fees for defending a negligence claim and any subsequent damages awarded.

Latent defects insurance

Latent defects insurance typically protects the owner against the cost of remedying the structure of a building, due to a defect. Usually, it lasts for 10 years from the original construction of a building. Typically, a building owner must arrange the cover in advance.

Other insurances

Product liability insurance. Product liability insurance protects against liability for injury to people or damage to property, arising out of products supplied by a business. Suppliers of equipment to a construction or engineering project, such as lifts or escalators, may be required to maintain such insurance, sometimes in place of professional indemnity insurance.

Adjoining Properties: Non-negligent liability insurance. Construction is an inherently dangerous process, particularly when it involves working on or near to existing or neighbouring buildings or other structures. No matter how much care is exercised, there is always the possibility that such property will suffer damage. Nobody has been negligent, but nevertheless, the owner has suffered a loss for which there is almost certainly no cover under their material damage insurance. Work involving any of the following is more likely to cause damage to adjoining properties:

- Demolition close to neighbouring property
- Excavation works near to existing foundations
- Piling
- Underpinning
- De-watering
- Shoring
- Work affecting the load-bearing capacity of an existing structure
- Work on listed buildings and buildings in poor condition.

This is a 'non-negligent' cover, and the exclusions are, therefore, important to help understand the intention of the wording. The main ones are:

- Damage caused by the negligence, omission or default of the contractor or any specialty contractor
- Damage attributable to error or omissions in the designing of the works
- Damage which can reasonably be foreseen to be inevitable having regard to the nature of the work or the manner of its execution.

Environmental liability insurance. This covers the cost of restoring damage caused by environmental accidents, such as pollution of land, water, air and biodiversity damage.

International insurance. There are three approaches: one policy to cover the parent company and all its interests worldwide; stand-alone policies in each country and a single global master policy issued to the parent combined with local policies in each country.

Wrap-up policies. Traditionally, each participant in a construction project obtains insurance individually to protect against the risk of financial loss. In recent years, 'wrap-up' insurance programmes have emerged as an alternative to the traditional method of risk management. In a wrap-up programme, the project owner can purchase an insurance policy that will cover the participants involved in the construction project, including the owner, construction manager, general contractors and specialty contractors. Typical wrap-up policies provide coverage for workers' compensation, general liability and builder's risk; however, programme features vary based on the insurance company and the type of project. The most common type of wrap-up programme is an owner-controlled insurance programme (OCIP). The same approach has been adopted more recently by construction managers and general contractors and is referred to as a contractor-controlled insurance programme (CCIP). Both programmes share the same key concept and many advantages and disadvantages.

Bonds, guarantees and warranties and third-party rights

Bonds, guarantees and warranties can offer some security from risks for both the client and the contractor. Breach of contract, where the contractor has failed to comply with its obligations to complete the works in accordance with the building contract, is one example of a risk.

Performance bonds allow the client to draw upon a secure fund, while parent company guarantees can be used in cases of insolvency to lower the risk. 'The Contract (Rights of Third Parties) Act 1999 provides further security to investors and can also be used as an alternative to collateral warranties by subcontractors. It enables third-party beneficiaries, such as investors and subcontractors, to enforce terms of a contract to which they themselves are not a party' (RICS, 2015a).

Bonds

A bond is a promise (usually by deed) whereby the person giving the promise (the bondsman) promises to pay another person (the employer) a sum of money. The bondsman only becomes obliged to make payment when called on to do so.

The contract will contain any details of bid or performance bonds required. There are different types of bonds:

- On-demand (simple) bond
- Performance bond
- Advance payment bond
- Off-site materials bond
- Bid bond (or tender bond)
- Retention bond
- Defects liability bond (or defects liability demand guarantee
- Adjudication bond.

On-demand (or simple) bonds require winning bidders to place a sum of money that can be collected by the client at its discretion, without having to prove through legal action that the contractor had defaulted on its contract. On-demand bonds are highly contentious.

A performance bond is a contractual undertaking by a bondsman to pay a specified amount to a named beneficiary on the occurrence of a certain event, usually the non-fulfilment of an obligation in an underlying contract. Performance bond costs are a small percentage of the full contract amount, which is often between 0.5% and 3% of the project cost. Most construction performance bonds are actually guarantees. Bonds and guarantees are related, but they are very different legal instruments. The right to claim under a guarantee is linked to non-performance of the underlying contract. Under a bond, the bonding organisation to pay is required to pay on demand regardless of the underlying contract.

Advance payment bond. If an advance payment to the contractor is agreed by the client, maybe to cover start-up costs, an advance payment bond may be required. This will secure the payment against default by the contractor and is usually an on-demand bond.

Off-site materials bonds are used where the client has paid for items, but they are not on-site, such as payment for goods ordered but not delivered. The bond, usually an on-demand one, will cover the cost of the items but will reduce as they are delivered.

Bid (or tender) bonds are often a requirement in an international tender process. They are usually on-demand bonds and are intended to secure the tender's commitment to commence the contract.

Retention bonds are used to ensure that the contractor undertakes and finishes the activities prescribed in the contract. They are based on a percentage (often 5%) of the contract amount, and the bond value will reduce after completion has been confirmed (Designing Buildings Wiki, 2015).

Guarantees

Guarantees include parent company guarantees or conditional bonds (secondary obligation instruments) where the bondsman is only liable where a breach of contract has occurred, for example, the contractor is in breach of contract. Due to their nature they are more common in the domestic construction market, and contractors are more likely to provide such forms of security than the on-demand variety (Designing Buildings Wiki, 2015).

Warranties

A warranty is an assurance that specific facts or conditions are true or will happen. The other party may rely on that assurance and seek a remedy if it is not true. Collateral warranties allow a 'duty of care' to be extended by the contractor to a third party who is not part of the original contract. 'They came into being as a result of the courts deciding that defects in buildings were not recoverable in tort, as they were an economic loss which was only recoverable through a contractual relationship' (Designing Buildings Wiki, 2015).

8.3 Site records

Amendment and updating. There are costs attached to completing, amending, communicating, sharing and storing site records such as people's time, storage facilities and back-up provisions. Accurate and up-to-date information is important for the day-to-day running of a project but can also be vital in any arbitration or court cases.

Responsible person. Site records should be maintained by designated (responsible) people such as the temporary works manager, the health and safety manager and so on.

Storage of paper copies of forms, reports, letters and so on would need to be in secure fire-proof storage units on-site. Electronic copies of records would need to be, on one hand, accessible to those who need the information, but on the other, security is important, especially with commercially-sensitive information.

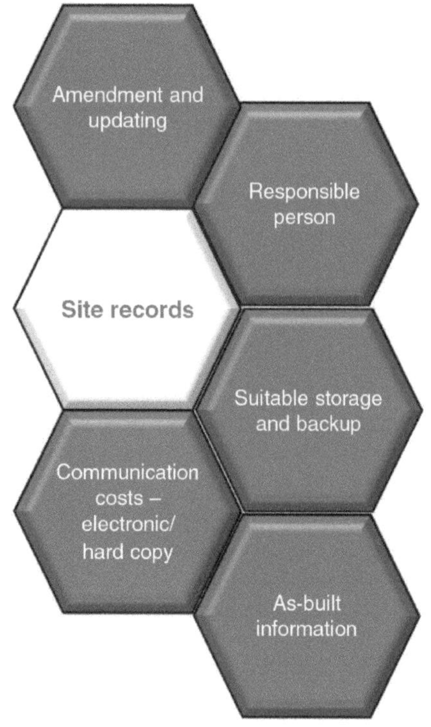

Back-up facilities would need to be provided off-site, using 'head office' facilities where available or using a cloud server provider. A cloud is an electronic structure where data is stored over many different computers and served up via a network connection, typically the Internet. It is particularly useful on a remote site and has the advantage of lower set-up costs than an in-house server.

As-built information. In the tender documents, the contractor may be required to produce 'as-built' drawings at the completion of the project to be handed to the client. The 'as-built' information is a record of what was produced, based upon the architect's and the engineers' design drawings. The contractor will mark up changes to the final construction issue drawings. It is important to capture the specialty contractors' record of what has been installed on-site, such as cabling information for the electrics. Such information is used in the maintenance and management of the project in use. The as-built drawings must reflect any changes made during the construction process. As-built information is not a standard service, unless specifically requested in the tender. The pricing of the production of the as-built information will be a time-and-resource lump sum.

Maintaining and communicating timely and accurate sources of information can help link different processes and different parties involved in the project. Therefore, assigning the responsibility for site records is important. Health and safety reports and risk assessments are a matter of regulation compliance and could incur penalties if not maintained.

8.4 Fees and charges

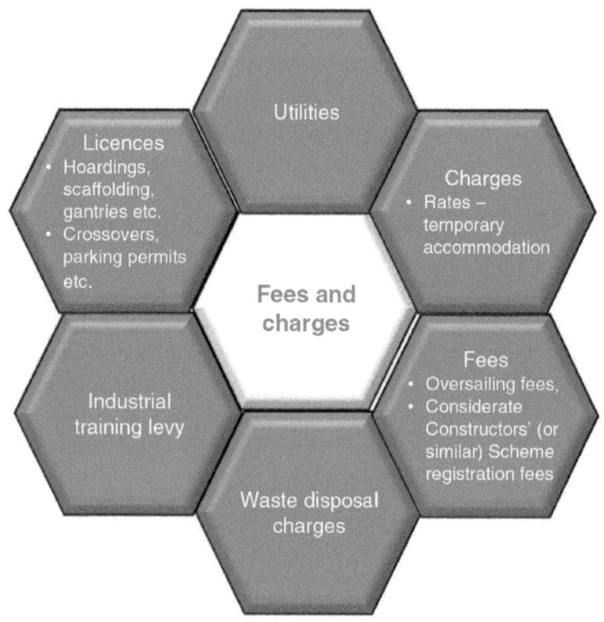

The honeycomb shows the considerations under Fees and charges.

Utilities fees and charges relate to connection costs and usage. This is explained in more detail under Temporary Services in the ***Preliminaries*** honeycomb.

The *Industrial Training Levy* is used in the United Kingdom (other countries have similar training levies) to provide training grants and other services that support the UK construction industry.

The Construction Industry Training Board (CITB) collects levies and distributes grants. Small companies, with a wage bill of less than £79,999, are not required to pay the levy. The contribution to the CITB appears in the ***Labour all-in rate*** honeycomb.

Waste disposal charges vary according to each local authority. Waste collection and disposal fees are usually based on weight. Minimising the weight through recycling, re-use and reduction (see Environmental Management honeycomb) can help keep those costs down. Assessing the waste likely to arise from a project is an important process at the estimating stage based on historical data from past projects or quotation from specialist waste disposal and management companies.

Other fees include those for oversail licences – either from the local authority (where the temporary structure, i.e. the use of a crane, crosses over the highway) or from the owner of an adjacent property. Local authorities may require certain documentation with the licence application, such as:

- Evidence of public liability insurance
- Site drawings
- The relevant risk assessment and reference to health and safety requirements
- The method statement dealing with construction, operation and dismantling of the crane
- Compliance with other relevant laws.

Scheme registration fees, such as the Considerate Constructors' Scheme, need to be allowed for in the estimate, where they are not paid for by the client.

8.5 Compliance

Compliance is an important issue for companies who are faced with a growing amount of regulation and legislation with the need to document and report many of the activities on-site. Off-site, financial regulations and corporate social responsibility (CSR) legislation are the responsibilities of the 'head office' or, in the case of an small- and medium-sized enterprise (SME), would probably be dealt with by the main contractor. The honeycomb overleaf shows the wide range of legislation associated with each of the areas.

The Considerate Constructors Scheme *was founded in 1997 by the UK construction industry to improve its image. Construction sites, companies and suppliers voluntarily register with the Scheme and agree to abide by the Code of Considerate Practice, designed to encourage best practice beyond statutory requirements. The Scheme is concerned about any area of construction activity that may have a direct or indirect impact on the image of the industry as a whole. The main areas of concern fall into three categories:*

- *The general public*
- *The workforce*
- *The environment*

The contractor has an obligation to ensure that work on the project complies with the relevant legislation. This necessitates communication with local authority officers (e.g. building control), factory inspector, health and safety inspectors and any other officials that need to check/supervise the way the work is being carried out. Inspectors may also need access to the site to check on work done by speciality contractors.

Site inspections might include (Designing Buildings Wiki, 2015):

- Planning inspections to verify compliance with planning permissions, conditions and obligations
- Inspections by representatives of funding bodies to review progress and quality for the release of money
- Inspections by insurers to ensure compliance with their terms and conditions
- Highways Authority inspection of roads and sewers to review any damage to roads and crossovers
- Health and Safety Executive inspections for compliance with health and safety standards
- Building control officer or approved inspector inspections during the progress of the work

Preliminaries

- Environmental Health Officer – inspections related to pollution (mud, noise, smoke and water) and certain installations, such as drainage and kitchens
- Fire Officer – inspection of fire escapes and for hazards, for example storage of certain materials
- Tree Preservation officer – inspection of protected trees
- Archaeological inspection of excavations, where the site is an archaeological area
- Factory inspectorate inspections.

The cost implication of compliance is the level of administration in completing forms and filing reports. The cost of time for meetings and inspections is also an issue. Health and safety is particularly heavily legislated; explained further in the section on Safety under Preliminaries. Labour laws, minimum wage legislation and so on have increased over time and have a big impact on the project processes, costs and duration. This will be discussed in further detail in the section on Staff under Preliminaries.

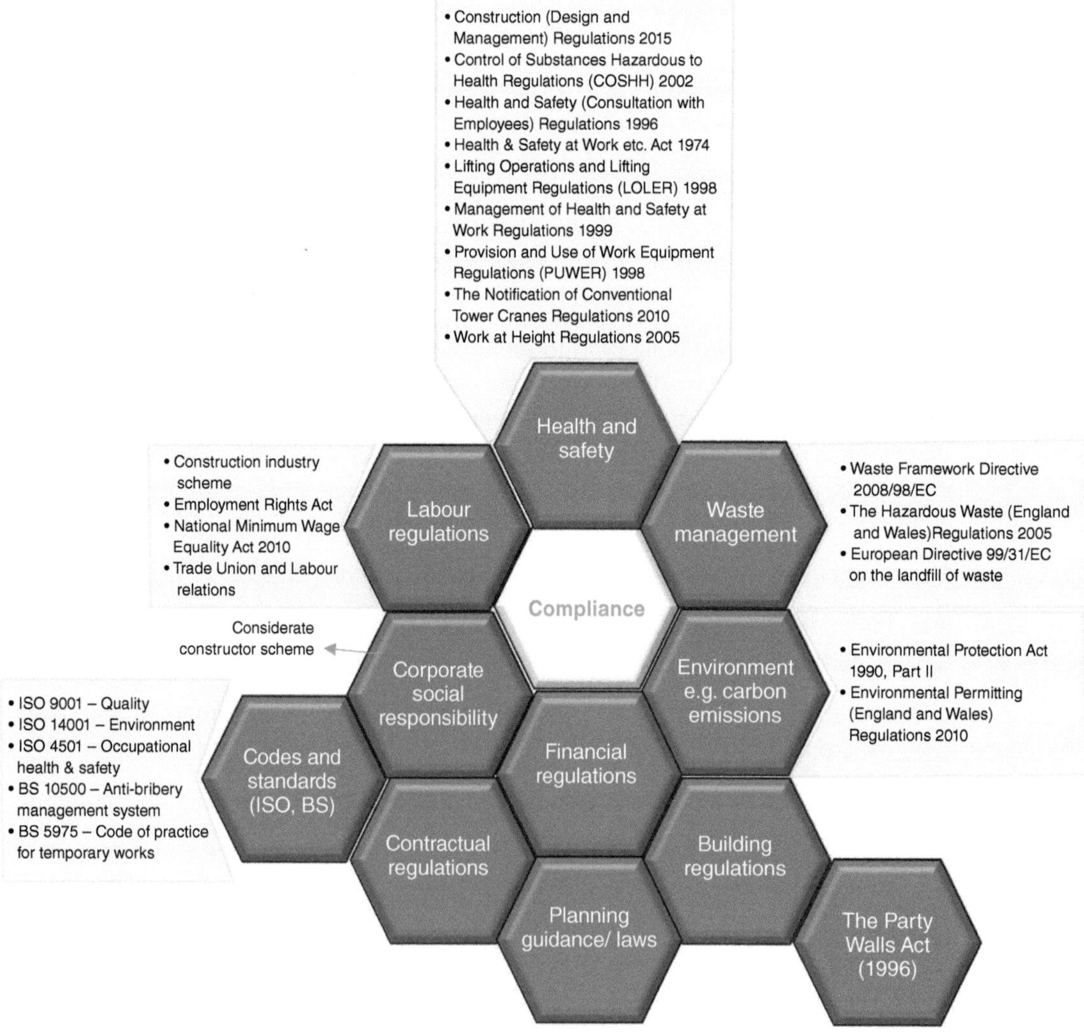

8.6 Environmental management

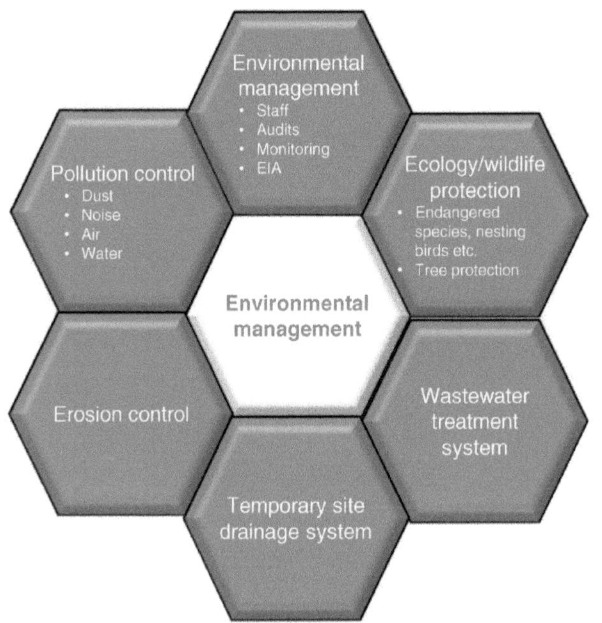

Environmental manager

This role may be an in-house expert or an external consultant in a large company. An environmental, or sustainability, manager oversees the environmental performance of the project. They develop, implement and monitor environmental strategies, policies and programmes that promote sustainable development. An environmental manager would also review the whole business, not just the project underway, resolving problems and introducing any necessary changes. Staff training and the communication of environmental information to all staff is another part of the role.

Environmental monitoring/audits

These are undertaken by the environmental manager or external consultant to analyse and report on environmental performance to both the company and external clients and regulatory bodies. It is an assessment of the extent to which an organisation is observing practices which minimise harm to the environment.

A compliance audit is an important part of the environmental audit process as there are cost implications for non-compliance (fines, fees, etc.) and there may be an impact on the company's corporate social responsibilities. Audits are particularly important if the client has stipulated sustainable practices on the project. Although companies are not legally obliged to have an Environmental Management System (EMS) in place, it is very advantageous in helping manage the company's environmental aspects.

Tree/ecology/wildlife protection

Being proactive in the conservation of wildlife can be cost efficient when compared to the cost of getting it wrong, and can improve the business case for the construction industry and the buildings and works it creates.
CIRIA (2011)

The protection may cover species (e.g. bats, badgers and great crested newts), sites (e.g. wetlands and nature reserves) and habitats and other species (trees, plants and wildlife). Ensuring this protection is important for a number of reasons:

- Legal – the law protects certain species
- Costs – delays if planning and legislative issues are ignored
- Good practice
- People/public opinion
- The penalties for damaging the environment can be heavy fines or even imprisonment.

Figure 8.4 shows the species, habitats of UK wildlife and where a survey licence is required. This calendar will have an impact on time and cost, depending on the species found on the site.

	When to survey (dependent on weather conditions)											
Species	**Jan**	**Feb**	**Mar**	**Apr**	**May**	**Jun**	**Jul**	**Aug**	**Sep**	**Oct**	**Nov**	**Dec**
Badgers												
Bats (hibernation roosts)												
Bats (summer roosts)												
Bats (foraging/commuting)												
Birds (breeding)												
Birds (winter behaviour)												
Dormice												
Great crested newts												
Invertebrates												
Natterjack toads												
Otters	Any time of year but better in summer as signs may get washed away in winter months											
Reptiles												
Water voles												
White-clawed crayfish												

Figure 8.4 Ecology species assessment calendar.

Temporary site drainage system

Construction work will inevitably involve run-off surface water into the water environment, and temporary drainage may be necessary on a site during the construction phase. The level and quality of that water is an important consideration in assessing the need (and cost) of constructing a temporary site drainage system. Sediment can be an issue in a drainage system during construction, particularly with the likelihood of cement or plaster wash-off into the drainage system.

A temporary sediment storage pond, which allows the sediment to settle, may need to be constructed.

Temporary storage of water may also be necessary in extreme conditions where flooding occurs. Areas for temporary storage can be car parking or recreational areas on large sites. CDM regulations cover safety aspects of the design, construction, maintenance and operation of drainage systems, temporary of otherwise. Run-off surface water may cause substantial soil erosion – see Erosion control.

Pollution control

Pollution can be considered as the introduction of a substance that has potential to cause harm to the environment or any organism supported by the environment. Pollutants may include, but are not limited to, silty waters, dust, mud, oils, chemicals and litter. The receiving environment might include air, ground, surface water and ground water. Key pollution risks from construction may arise from any of the following activities (*Source: CIOB Carbon Action 2050, Wielebski (2013))*:

- Site set up, welfare facilities and management
- Earthworks and foundations
- Tunnelling works
- Stockpiling

- Site drainage
- Working near underground services
- Working in or near watercourses
- Storage of fuel, oils and chemicals
- Bridge cleaning and repainting
- Plant and wheel washing
- Pouring and handling of concrete
- Refuelling of plant and equipment
- Topsoil stripping
- Waste and material handling, storage and disposal
- Silt from excavations, exposed ground, stockpiles and site roads
- Unauthorised discharges to surface waters.

Noise

Noise is considered a 'statutory nuisance' if it is 'prejudicial to health or a nuisance' (Environmental Protection Act, 1990, 79 (1)(g)). Noise is inevitable on a construction site, and operations may have conditions placed upon them by the local planning authority (National Planning Policy Framework, 2012). Noise can be mitigated in a number of ways: reducing noise at source (engineered solution); changing layouts and/or use of screening and noise insulation. Enclosing noisy equipment, such as a generator, is one option.

Noise is also an issue in the welfare of the site personnel (The Control of Noise at Work Regulations, 2005). Noise at work can interfere with communications and make warnings harder to hear. It can also reduce people's awareness of their surroundings. These issues can lead to safety risks – putting people at risk of injury or death. The duties under the Regulations include the need to:

- Ensure that the legal limits on noise exposure are not exceeded;
- Maintain and ensure the use of equipment to control noise risks;
- Provide employees with information, instruction and training and
- Undertake health surveillance (monitor workers' hearing ability).

Air and ground pollution

Volatile organic compounds (VOCs), such as formaldehyde and benzene, may be released as vapours from fuels, petroleum solvents and bituminous tar oils or leave deposits polluting the soil and groundwater. The Method statement should include methods that minimise pollution emissions. Materials used in construction operations such as oil, fuel, lubricants, chemicals, cement, lime, paint, cleaning materials and others have the potential to cause serious pollution impacts if not correctly managed.

Dust

High levels of dust are created in a number of construction procedures. These can be hazardous to workers' health and a nuisance to neighbouring properties. Sawing, sanding, grinding, blasting and sweeping can create a lot of dust. There are a number of ways of limiting the dust and/or protecting against it: avoiding cutting, using a less powerful tool, or a different work method. Water helps to damp down dust clouds, and vacuum extraction can be used. The aim is to, in the first instance, prevent pollution and then contain it where possible; measures that can be taken include:

- All CoSHH materials must be stored in secure bunded areas
- Fuel/oil stores must be located away from site drainage systems and watercourses
- Any tanks or drums need to be stored in a secure container or compound
- The contents of any tank need to be clearly marked on the tank
- A Material Safety Data Sheet (MSDS) should be kept on file for each material used on-site
- An inventory of all COSHH materials should be maintained by the store supervisor.

8.7 Wastewater treatment system

Untreated wastewater can cause major pollution both in the vicinity of the site and beyond. Therefore, wastewater treatment system is important. When drawing up wastewater treatment proposals for any development, the first presumption is to provide a system of foul drainage discharging into a public sewer to be treated at a public sewage treatment works (those provided and operated by the water and sewerage companies). This should be done in consultation with the sewerage company of the area. Where a connection to a public sewage treatment plant is not feasible (in terms of cost and/or practicality), a package sewage treatment plant can be considered. This could either be adopted in due course by the sewerage company or owned and operated under a new appointment or variation. The package sewage treatment plant should offer treatment so that the final discharge from it meets the standards set by the appropriate agency.

Vehicle-washing bay

Keeping mud and dust off the roads is a requirement on many sites. Using vehicle-washing bays ensures minimum disruption to roads and footways. Surface run-off from washing areas can contain high levels of pollutants, for example:

- Detergents
- Oil and fuel
- Suspended solids
- Grease
- Antifreeze.

The run-off must not be allowed to enter surface water drains, surface waters or ground waters as it will cause pollution and the offender prosecuted. Vehicles should only be washed in defined areas where the wash water and any rainfall run-off can be contained. If possible, the surface run-off from the vehicle washing area should be directed to an on-site treatment system. These guidelines apply to all contractors/organisations using the site and are the main contractor's responsibility.

Erosion control

Run-off water from site activities can cause soil erosion, and so appropriate treatment of it can help to avoid erosion. Stockpiles of soil on the site should be seeded with a grass/clover mix to minimise soil erosion and to help reduce infestation by nuisance weeds that might spread seed onto adjacent land.

8.8 Waste management

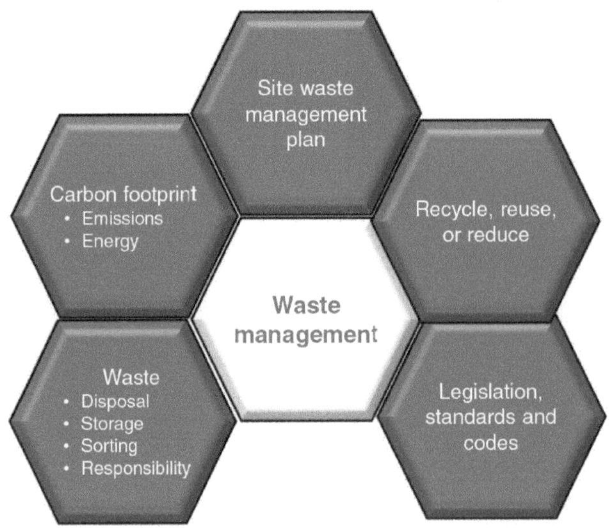

Effective waste management begins at the design stage. Well-considered design solutions, for example plasterboard sizes that have synergy with the height of internal walls and stud partitions can have a significant impact in terms of reduced waste. Moreover, the emerging role of Business Information Modelling (BIM) has the propensity to introduce a raft of benefits, for example improved construction detailing that can negate abortive and wasteful re-work to the availability of more precise quantities (CIOB, 2013). All waste produced can present a safety hazard to workers on-site if it is not properly managed throughout the project. Decisions need to be made at an early stage on:

- How – wastes streams produced during building work will be managed in a timely and effective way;
- Who – is responsible for collecting and disposal of specific waste produced on-site. Problems often arise when company and individual duties are not made clear before work starts.

Site waste management plan

Until 2013, a site waste management plan (SWMP) – see Fig. 8.5 – was a legal requirement in the United Kingdom for projects over £30,000. Since then, the creation of an SWMP has become a matter of good practice which helps to reduce waste (and therefore the cost of disposal) and improve environmental performance. Having a waste management plan (WMP), and thus achieving lower waste allowances, can help to reduce a contractor's bid, as well as showing a commitment to the environment. The UK Building Research Establishment (BRE) has a developed a SMART Waste plan made up of 3 stages (Fig. 8.6).

The true cost of waste

The Landfill Tax is charged by weight. The standard rate is £84.40 per tonne; inert or inactive waste is charged at £2.65 per tonne.

The true cost of waste is not just the cost of hiring a skip. It also includes:

- Cost of the materials that end up as waste
- Labour cost to handle the waste on-site
- Cost of waste storage, transport, treatment and disposal.

Preliminaries

The true cost of construction waste will continue to rise substantially each year due to:

- Landfill Tax increasing each year (£86.10 per tonne in 2017)
- Higher disposal charges
- Purchase costs of materials and products increasing.

Site Waste Management Plan data sheet			
Project name:			
Date when this sheet was filled out:			
Stage of project (eg planning stage, during project delivery, end of project):			
Report number (projected waste arising should be report number one etc):			
Project address / location:			
Estimated cost of the project:			
Client:			
Principal contractor:			
Person responsible for waste management on site (name and job title):			
Person and company completing this form, if different:			
Sites your waste is going to (including permit, licence or registered exemption reference number and details):			
A	B	C	D (add more boxes if needed)
Details of the people removing waste from your site (including their waste carrier registration number):			
A	B	C	D (add more boxes if needed)

		Quantity (specify volume or weight, eg m³, kg, T, number of skips)																
Types of waste arising (add more rows if needed):	**EWC** code**	**Reused** on site		off site		**Recycled** for use on site		for use off site		Sent to recycling or reprocessing facility		**Disposed of** land-fill		other than landfill (eg incinerated)		Relate to boxes above (ie insert A or B etc)		**WTN † completed?**
Target / Achieved (T) / (A)		T	A	T	A	T	A	T	A	T	A	T	A	T	A	Waste site	Waste carrier	
Inert																		
Non-hazardous																		
Hazardous																		
Totals (m³, kg, T)																		
Performance score as %*																		
SWMP Target %*																		

Figure 8.5 An example of an SWMP data sheet.

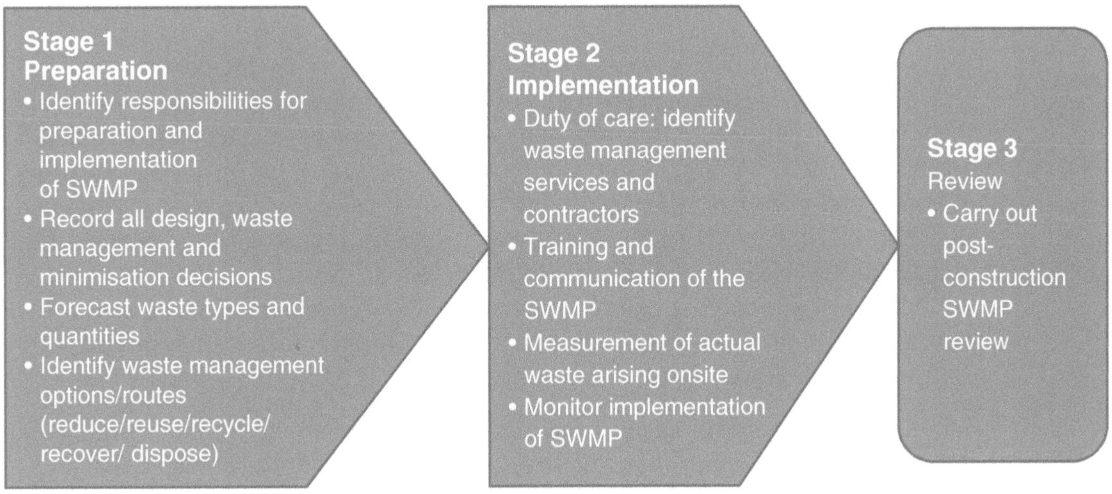

Figure 8.6 The three stages of the SMART Waste plan.

8.9 Waste disposal, sorting and storage

Waste disposal

A business has a legal responsibility to ensure that they produce, store, transport and dispose of business waste without harming the environment. This is called a 'duty of care'. The duty of care has no time limit. A contractor is responsible for their waste from the time of production until it is transferred to an authorised person. The duty of care does not end when the waste is handed over; it extends along the entire chain of management of the waste. A contractor should:

- Segregate, store and transport waste appropriately and securely, making sure that it does not cause any pollution or harm to human health;
- Check that the waste is transported and handled by people or businesses that are authorised to do so and
- Complete waste transfer notes, including a full, accurate description of the waste, to document all waste transferred and keep them as a record for at least 2 years.

Commercial waste needs to be classified before it is disposed. The UK List of Waste (LoW) classification can be used; it includes:

17.01	Concrete, bricks, tiles and ceramics
17.02	Wood, glass and plastic
17.03	Bituminous mixtures, coal tar and tarred products
17.04	Metals (including their alloys)
17.05	Soil (including excavated soil from contaminated sites), stones and dredging spoil
17.06	Insulation materials and asbestos-containing construction materials
17.08	Gypsum-based construction material
17.09	Other construction and demolition wastes.

An assessment needs to be made as to whether or not the material is hazardous. This can be ascertained from the manufacturer's safety data sheet. All of this information is needed to complete waste documents and records.

On-site sorting and storage

Skips are the most common method of waste collection. These are delivered and collected by licensed waste management firms. Tipping skips, or mini-skips, can help in the handling of site waste. They are available with capacities ranging from 0.25 to 2.5 cubic yards (0.2–1.9 m^3). Larger models reach capacities of up to 6 cubic yards (4.6 m^3). Integral forklift channels on the base of the skips enable easy pick up and transport using a forklift truck. Separating the waste at source can be achieved by using different colour skips, making it easier to recycle. Signage and education of the workforce is important to re-enforce the importance of green waste management. The positioning of the skips is also important to ensure that distances from source to skip are not too great.

Sorting bricks, tiles and wood for re-use has a financial and environmental gain, although there is a labour cost and storage issues involved. Inert materials such as concrete, brick, asphalt, soils and stones can be reused on-site as hardcore or for backfill at other excavation sites. Topsoil can be reused for landscaping or as part of compost once necessary tests have been carried out (NSCC, 2007).

Carbon footprint

A carbon footprint of a building can be defined as the carbon dioxide (CO_2) emissions resulting from construction materials, construction activities, lifespan operation and eventual demolition. A carbon footprint can also be expressed in carbon dioxide equivalent (CO_{2e}), which is a measure of how much global warming a given quantity of greenhouse gas may cause using CO_2 as a reference. The term 'carbon' is commonly used when referring generically to either CO_2 or CO_{2e} emissions.

There are several choices related to the carbon footprint at the bid stage that will impact cost estimating:

- The choice of local suppliers to reduce transport-related emissions
- More energy-efficient plant and machinery
- Low-carbon construction materials
- Good waste management, including recycling, re-use and reduce.

Recycle, re-use or reduce

Recycled materials used in construction:

- *glass*
- *paper*
- *plasterboard*
- *compost and other organics*
- *plastics*
- *rubber*
- *wood*
- *aggregates*
- *cement replacement – for example pulverised fuel ash*

The 3 Rs are concerned with better resource efficiency in accordance with the following principles:

- Reduce – eliminating the generation of waste, where possible, by stopping it coming on to site in the first place
- Reuse – making use of materials in their original state on the same site or at other sites
- Recycle – turning materials into new products for other purposes.

The final stage in this process, which completes the 3 Rs 'loop', is the specification and use of materials with higher recycled content on future projects to further reduce the demand on natural resources (NSCC, 2007). There are many business benefits in using recycled materials on a construction project:

- Reducing material and waste disposal costs
- Increasing competitive advantage

Table 8.1 Example of a checklist for WRAP's Net Waste Tool.

	Information required	Information source document	Person responsible
Step 1	Project		
	Project timeline		
	Costs		
Step 2	Estimated waste sources and quantities		
	Wastage rates		
	Proposed waste reduction actions		
Step 3	Waste segregation		
	Waste container types and rates		
Step 4	Waste recovery – target recovery rates and actions required		
Step 5	Transfer Quick Wins to the site waste management plan		

- Reducing CO_2 emissions
- Meeting your planning requirements
- Complementing other aspects of eco-design
- Responding to and pre-empting changes in public policy, such as increases in Landfill Tax
- Responding to client requirements.

WRAP

The Waste and Resources Action Programme (WRAP) works with UK government bodies, the European Union and other funders to help deliver their policies on waste prevention and resource efficiency. They are a registered charity and provide a number of tools for businesses to help reduce, recycle and re-use materials. The Net Waste Tool (see Table 8.1) is one of these; it helps to reduce the amount of waste going to landfill.

The tool enables:

- Identification of opportunities to increase recycled and reused content in materials
- Project-level and corporate reporting
- Forecast waste to be compared with actual
- Calculation of metrics including value wasted, disposal cost, waste reduction and diversion of waste from landfill and is applicable to a wide range of construction projects – from a small to medium to mega developments.

Legislation/ guidance includes:

- *The Waste (England and Wales) (Amendment) Regulations 2012*
- *EU Waste Framework Directive*
- *Environmental Protection Act 1990*
- *The Hazardous Waste (England and Wales) Regulations (S.I.2005/894)*
- *Approved Document H (Drainage and waste disposal)*
- *Planning Policy Statement 10: Planning for sustainable waste management*

Legislation, standards and codes

The EU Waste Framework Directive provides the legislative framework for the collection, transport, recovery and disposal of waste for its members. The UK regulations are based on this Directive. The Directive requires all member states to take the necessary measures to ensure that waste is recovered or disposed of without endangering human health or causing harm to the environment and includes permitting, registration and inspection requirements.

8.10 Setting out

Setting out can be a high risk process as mistakes at this stage can lead to expensive corrective work and delays. Following on from the site clearance, setting out marks the position of the foundation trenches. The increasing use of GPS instruments has helped in the process both with levels and heights in the setting out process. The setting out process involves the client, the engineer and the contractor. Whilst the engineer checks the process, it is the responsibility of the contractor.

The method of setting out is the reverse of surveying process. The process involves the positions and levels of building lines and road alignments shown on the construction plans to be established on the ground by various techniques and instruments.
Civil Engineering Dictionary (2015)

Site levels need to be confirmed before setting out any earthworks, a process that could be supported by a digital ground model. Setting out information is provided by the designers but may not be included in the tender document drawings, so timing is an issue estimating setting out costs. On large complex sites, the decision may be to **outsource** the work to specialists; this will have a cost and time implication. If carried out by the contractor, **surveying tools and equipment** (see Table 8.2) will be needed, either ones they already have or it may be necessary to hire, especially for a specialist item.

The choice of which tools should be used will depend on a number of factors (Sadgrove, 2007):

- Size of the site
- Complexity of the work
- Precision/accuracy demanded
- Economics: the time a task requires may be a dominating factor.

Primary grids, such as a survey grid or a site grid (used by the designer), enable points to be set over a large area.

Topographic surveys provide more complex information about the existing site. Topographic information can be obtained through digital maps, or the survey can be undertaken by specialists.

Table 8.2 Surveying and setting out tools and equipment.

Type	
Theodolite	Optical/manual (analogue) instrument for the measurement of angles only Most common electronic instrument used on site for the measurement of both angles and distances. Measurement information displayed digitally can be stored in a data logger.
Total Station	The Total Station includes a built-in computer that provides the site engineer with a number of features that typically include ■ input of ambient metrological conditions ■ observed positions as Cartesian coordinates or bearing and distances ■ distances corrected for slope ■ setting-out mode either using either coordinates, offsets or bearing and distances. Reflectorless Total Stations do not need a prism reflector set at the target. Robotic Total Stations do not need an operator at the instrument.
Gyroscopic theodolite	For measuring and setting-out angles relative to True North – especially useful when working underground
GPS	Generic term for a range of positioning and setting-out solutions using the US DoD's NavStar Global Positioning System (GPS) satellites. For the site engineer, the use of GPS for control is the most relevant application. Many GPSs now also incorporate the GLONASS satellite navigation system. NB: GPS-aided/robotic systems are becoming increasingly common for the automatic control of earth moving machinery.
Optical level	Optical/manual instrument only. Suitable for most site applications
Automatic level	Similar to the optical level in principle but susceptible to vibration
Digital level	Used with a 'bar-coded' staff for precise levelling
Precise level	Also known as a geodetic level. Only for very high accuracy control and requires a competent specialist operator
Optical plumb	Optical/manual instrument only. Average of four readings should be taken with instrument turned horizontally 90° between readings. For automatic version, normally two readings separated by 180° are sufficient
Lasers	Alignment – used to define a line/direction. Rotating (horizontal) – defines a horizontal plane. Rotating (general) – defines any set plane. Pipe – defines line and grade
Optical square	For setting-out right angles over short distances only

Laser scanning can be used to collect surface data; this is often called a point cloud survey. CAD and BIM systems allow the import of point cloud data into 3D visual graphic material.

Site information and field documents need to be completed and stored.

Soil testing will be needed for sites where there has been soil contamination. It may be done 'in-house' or outsourced. A soil survey provides details of the characteristics of soil on the site enabling an understanding of soil types and how they behave in different conditions (Designing Buildings Wiki, 2015).

8.11 Control and protection

This section of the cost considerations within the preliminaries includes four specific areas, as shown in the honeycomb.

Environmental control of building

- Dry out building
- Temporary heating/cooling

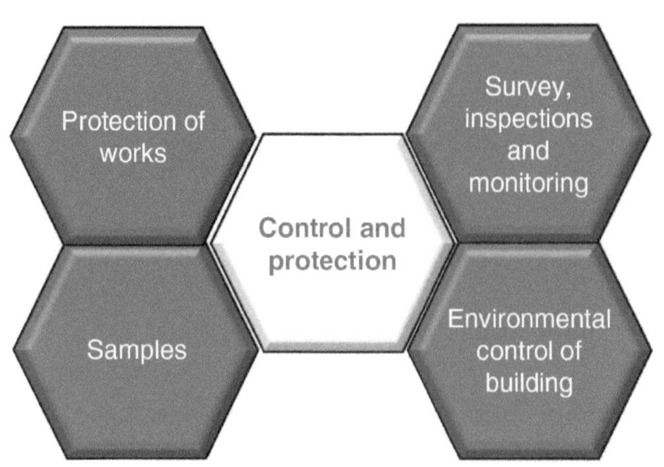

- Temporary waterproofing, including over roofs
- Temporary enclosures.

Protection of works

- Protection of finished works to project handover
- Protection of stairs, balustrades and the like works to project handover
- Protection of fittings and furnishings works to project handover
- Protection of entrance doors and frames works to project handover
- Protection of lift cars and doors works to project handover
- Protection of specifically vulnerable products to project handover
- Protection of all sundry items.

Survey, inspections and monitoring

- Surveys
- Topographical survey
- Non-employer dilapidation
- Structural/dilapidations survey adjoining buildings
- Environmental surveys
- Movement monitoring
- Maintenance and inspection costs.

Samples

- Provision of samples
- Provision of sample room
- Mock-ups and sample panels
- Testing of samples/mock-ups, including testing fees
- On-site laboratory equipment
- Mock-ups of prefabricated units (e.g. residential units, student accommodation units and hotels).

8.12 Completion and post-completion requirements

Once the contract is complete, handover takes place, which involves a number of requirements. A handover plan, agreed with the client, should set out the processes and documents involved.

During handover, the client should be issued with (Designing Buildings Wiki, 2015):

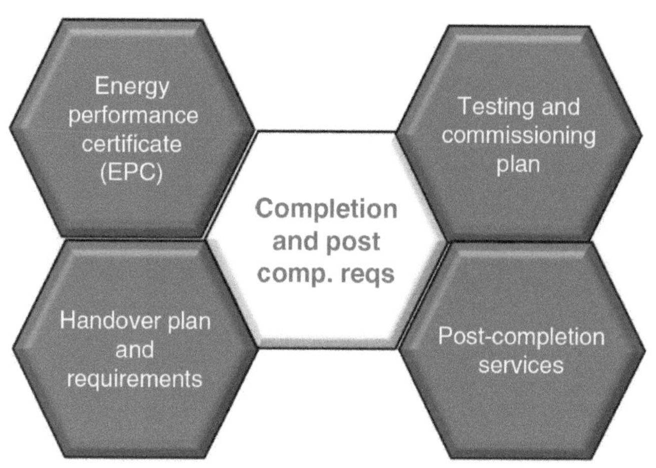

- Keys, fobs and transmitter controls for the development
- The health and safety file
- The draft building owner's manual
- The building log book, including as-built drawings and technical information
- A building user's guide
- Up-to-date testing and commissioning data
- All certificates and warranties in respect of the works
- As-built drawings from consultants and specialist suppliers and contractors (or as manufactured and installed) or an as-constructed building information model
- Copies of statutory approvals, waivers, consents and conditions
- Equipment test certificates for lifts, escalators, lifting equipment, cradle systems, boilers and pressure vessels
- Licences such as licences to store chemicals and gases and to extract groundwater from an artesian well.

Handover requirements include the training of building user's staff in the operation and maintenance of the building engineering services systems; the provision of spare parts for maintenance of building engineering services and the provision of tools and portable indicating instruments for the operation and maintenance of building engineering services systems. Certain services are required during the defect liability period (or other specified period) such as the operation and maintenance of building engineering services installations, mechanical plant and equipment.

A testing and commissioning plan needs to be produced with regard to building services.

Post-completion services involve the provision of staff, labour and materials to deal with defects (in the defect liability (or other specified) period) and may include the maintenance of internal and external planting.

An *Energy Performance certificate* (EPC) is required once the building has been constructed. It rates the energy efficiency of the building using grades from A to G ('A' is the most efficient grade). The failure to produce an EPC can result in a fine based on the rateable value of the building.

8.13 Contingencies

There are three basic types of contingencies in projects: tolerance in the specification, float in the schedule and money in the budget (CIRIA, 1996). The amount included for contingencies relates to the level of risk, which reduces over time as more information is available in the production process. 'At the preliminary business plan stage total cost estimates might include a 15% contingency; in the elemental cost plan this might reduce to 10% of fees and construction costs and on awarding

Table 8.3 Contingency estimating methods.

Contingency estimating methods	Description
Traditional percentage	Based on most likely value, the point estimates are calculated for each cost element. These are a project-wide percentage to the base estimate using historical information as well as the estimator's knowledge, experience and instincts.
Expected value	Individual risk occurrences have an 'expected value' that is: impact × probability of occurrence, which need to be calculated.
Method of moments	This method involves the calculation of a probability distribution for each cost item. The distribution reflects the risk. This is an extension of the expected value method.
Monte Carlo simulation	This computer-based method produces risk profiles using quantitative techniques. This allows a structured approach to setting the contingency value for the project.

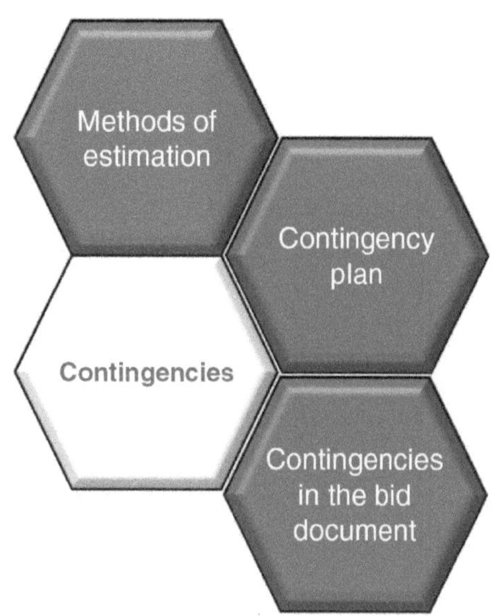

the contract, 5% of the contract value might be included as contingency in the cost plan' (Designing Buildings Wiki, 2015).

Contingency allowances can be referred to as a 'risk mark-up' but should not be confused with a risk allowance. Allowances are not risk-based but are used for expected events within the scope of the project. The client will specify the amount to be included for contingencies.

Methods of estimation

Estimating contingencies can be divided into two methods: deterministic (e.g. the traditional percentage method) and probabilistic (e.g. method of moments). Table 8.3 shows the various contingency estimating methods.

Contingency plan

Contingency plans are put in place to deal with the identified project risks if/when they occur. These risks include non-delivery of materials; lack of available work force; lack of plant and equipment availability (or plant failure or access to spare parts); illness; the risk of obtaining timely planning permission or other statutory approvals and information and communications technology failure (Designing Buildings Wiki, 2015).

8.14 Management and staff

The honeycomb (Fig. 8.7) shows the needs to be taken into consideration in estimating management and staff costs.

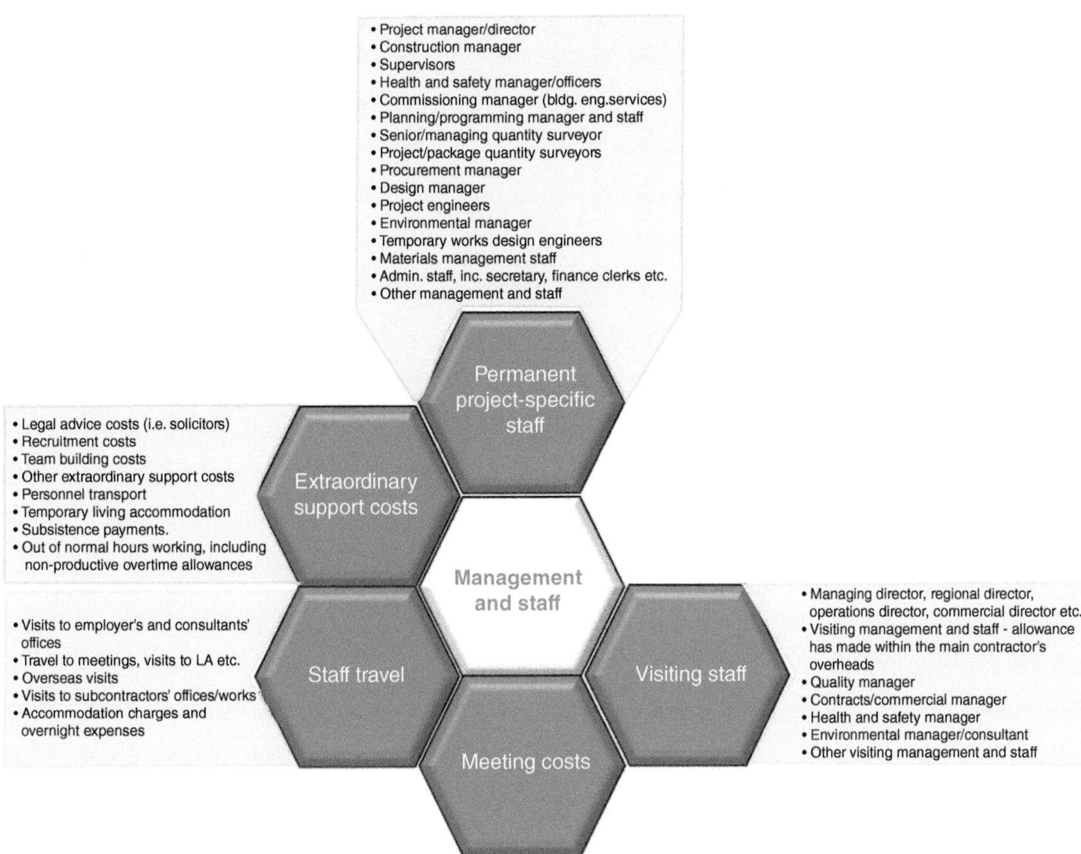

Figure 8.7 The management and staff honeycomb.

9 Temporary Works

9.1 Introduction

The term 'temporary' belies the importance of this part of a construction project. Whilst the works are removed by the end of the project, their design, execution and management are vitally important in terms of engineering, safety and efficiency. Time, cost and quality are all affected by temporary works (TWs). Examples of temporary works include, but are not limited to:

- Earthworks – trenches, excavations, temporary slopes and stockpiles.
- Structures – formwork, falsework, propping, façade retention, needling, shoring, edge protection, scaffolding, temporary bridges, site hoardings and signage, site fencing and cofferdams.
- Equipment/plant foundations – tower crane bases, supports, anchors and ties for construction hoists and mast climbing work platforms (MCWPs) and groundworks to provide suitable locations for plant erection, for example mobile cranes and piling rigs.

The nature of the work, such as scaffolding and excavation support, means that failure in any part of temporary works can often lead to a high-consequence/risk event that would be reportable in the United Kingdom under the Reporting of Injuries, Diseases and Dangerous Occurrences Regulations (RIDDOR) or the relevant health and safety regulations overseas.

The 'cost' can be much higher in terms of health and safety and reputation if the work is not carried out correctly. Table 9.1 shows temporary works classified (by the UK Health and Safety Executive (HSE)) as low, medium or high risk. The Bragg report, published in 1975, led to the development of a British Standard – BS 5975 1982 – the precursor to today's standard BS5975:2008+A1 2011 (see Codes and Guides). The Bragg Report maintained that falsework (which applies equally to all temporary works) needs as much attention to detail as the design of more permanent structures. Therefore, any falsework should be regarded as a structure that requires skill and competence in its construction as its stability forms the basis of subsequent work and is vital to safety (Carpenter, 2012).

The honeycomb in Fig. 9.1 shows the items involved in temporary works. These are described next in more detail.

New Code of Estimating Practice, First Edition. The Chartered Institute of Building.

Table 9.1 The different levels of risk in temporary works.

Design complexity risk	Type of design	Example
DCROO	Manageable design as per industry safe standard (CFRO/CFR1)	Small-scale groundwork, for example gentle slope
DCRO	Standard solution with minimal consequence of failure (CFRO/CFR1)	Standard scaffold; formwork/falsewok <3 m high
DCR1	Simple design with minimal consequence of failure	Pits and trenches to CIRIA 97 Trenching Practice Mobile crane outrigger Foundations in good ground crane to 50T
DCR2	More complex design with failure risk GFR2 or below	Special designed scaffolds Large, complex or unusual system formwork or falsework Cofferdams and sheet piled walls, contiguous piled temporary walls
DCR3	Complex/innovative designs – not CFR3	Bridge demolition Structurally complex partial demolition or modification of existing structures Trenchless construction Excavations and cofferdams in tidal conditions
DCR4	Abnormal/innovative design with a CFR3 category	Use of glass as a structural material Long span bridge erection

NB CFR = category of failure risk (for more information see Clients' guide to temporary works – TWf2014: 02 (temporary works forum)). CFRO – benign; CFR1 – low impact; CFR2 – potential major effect; CFR3 – catastrophic failure.
Source: HSE (2010) and Adapted from BS 5975.

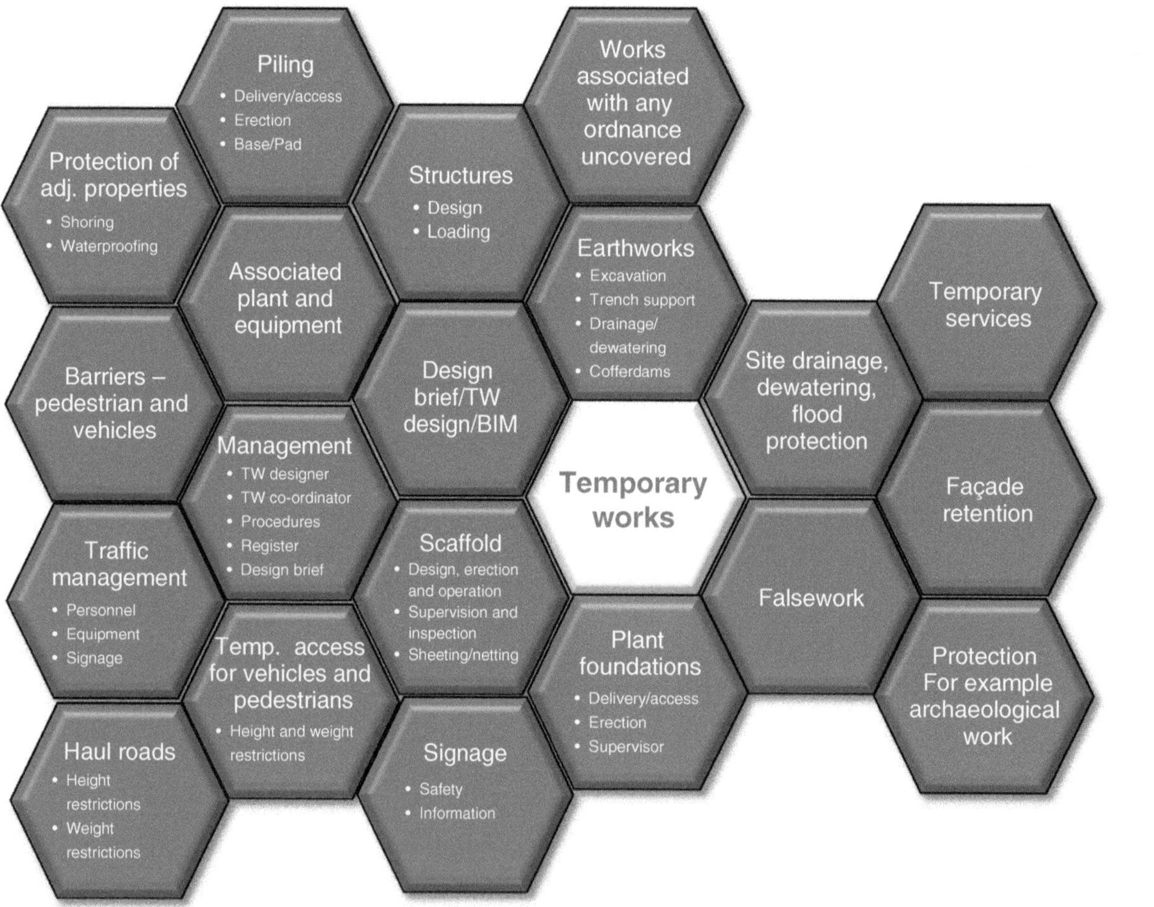

Figure 9.1 The temporary works honeycomb.

9.2 Temporary works management

The correct design and execution of temporary works is an essential element of risk prevention and mitigation in construction. The Management of Temporary Works, produced by the UK HSE, is designed to:

- Promote awareness and knowledge of the importance of managing TWs;
- Improve contractors' management arrangements of TW;
- Increase the competence of those engaged in TW management and design and reduce accidents arising from TW failures.

Contractors should be able to demonstrate that they have in place effective arrangements for controlling risks arising from the use of temporary works. These are usually captured in a TW procedure which will contain most or all of the following elements:

- Appointment of a temporary works co-ordinator (TWC)
- Preparation of an adequate design brief
- Completion and maintenance of a temporary works register
- Production of a temporary works design (including a design risk assessment and a designer's method statement where appropriate)
- Independent checking of the temporary works design
- Issue of a design/design check certificate, if appropriate
- Pre-erection inspection of the temporary works materials and components.

Control and supervision of the erection, safe use, maintenance and dismantling of the temporary works, that is, procedures to:

- Check that the temporary works have been erected in accordance with the design and issue a formal 'permit to load' where necessary
- Confirm when the permanent works have attained adequate strength to allow dismantling of the temporary works and issue a formal 'permit to dismantle' where necessary

The procedure should include measures to ensure that the design function and the role of TWC, and temporary works supervisor(s) where appropriate, are carried out by competent individuals

Smaller contractors may not have the experience to operate their own temporary works procedure and may need to obtain external expertise. It is also common for large and medium contractors to outsource aspects of temporary works design and management.

9.3 Temporary works co-ordinator (TWC)

Every organisation involved in temporary works should have a designated individual responsible for:

- Establishing, implementing and maintaining a procedure for the control of temporary works for that organisation and
- Ensuring that sub-contractors have adequate temporary works procedures if they are carrying out and managing temporary works.

The coordinator is not normally the designer and is responsible for ensuring that a suitable temporary works design is prepared, checked and implemented on-site in accordance with the relevant drawings and specification. BS5975: 2008, Clause 7.2.5 lists the principal activities of the TWC. The TWC needs to have a level of authority that allows them to stop the work if it is not satisfactory. Ideally, a TWC would:

- Have experience of the relevant types of temporary works.
- Have completed formal TWC training.
- Hold a degree/HND in civil/structural engineering.
- Be a chartered civil/structural engineer.

9.4 Temporary works supervisor

On larger sites where there are a number of sub- and specialty contractors, the TWC may need help in the supervision of temporary works. The supervisor(s) is/are the TWC's 'eyes and ears'.

9.5 Temporary works register

A temporary works register (see below) is a good reference document on any project and comprises a list of all the temporary works items identified for the project. These can be set out as a table using appropriate headings, which could include:

- Design brief number (for each item) and date issued
- Short description of temporary works
- Date required
- Category of temporary works
- Designer
- Design checker
- Date design complete
- Date design checked/approved
- Erection complete and checked or 'permit to load' and 'permit to dismantle'.

9.6 Temporary works design brief

The design brief should include all relevant information to enable an effective but economic design to be provided by the engineer including both underground and aboveground features that need to be taken into account within the design. A design brief should be prepared to serve as the starting point for subsequent decisions, design work, calculations and drawings. All concerned with the construction should contribute towards the preparation of the brief, which is based on the TW register. Figure 9.2 is an example of a TW design brief. It is important that the brief is prepared early enough to allow sufficient time for all subsequent activities, that is design, design check, procurement of equipment and construction/erection of the scheme.

Temporary works design

Temporary works need to be designed to the same high standard as permanent works in accordance with recognised engineering principles. A temporary works designer works closely with the TWC giving advice not only on safe and efficient temporary works but also (but not limited to) on construction methods/sequencing,

Temporary Works Design Brief | *No.:*

DB ref.	**Date**	**Project**	**Contract** Name and No.	**TW Reference**	**Contact**

Design ***A TW Design Certificate must be submitted, highlighting critical items for checking on site, risks and element not included in the design***			
Structure	Structure element	Design check category (0,1 2, 3)	Description

ITEM	**Attached**	**Part of design Y/N**	**N/A**
A Risk Information			
B Drawing References			
C Relevant Site/Ground Investigation			
D Existing Ground Conditions			
E Relevant Topographical Information			
F Existing Services Information – overhead & below ground			
G Traffic Management Requirements			
H Loading Criteria			
I Railway Interface			
J Interface with other Buildings or Structures, (& if listed)			
K Preferred Materials			
L Preferred Method – including preliminary sketches			
M Construction Phasing			
N Other Constraints/limitations/contract requirements			

Compiled by	Approved by	**Key dates**		
		Date submitted to designer	Design required by	Notice – working days
Date:	Date:			

Figure 9.2 An example of a TW design brief.

materials and construction method information, design risk assessment and contributing to the preparation of detailed method statements.

There are some key differences between temporary and permanent works of which the TW designer must be aware (Carpenter, 2012):

- Re-use of material
- Possibility of abuse on-site
- Lesser certainty of vertical accuracy in particular
- Some categories, for example falsework liable to receive full design load (unlike many permanent works)
- Essential need for robust lateral support to aboveground structures
- Complex dismantling scenarios, for example lack of space or underload.

The design needs to be checked for 'concept, adequacy, correctness and compliance with the requirements of the design brief' (BSI, 2011). Within the code, temporary works are further classified, for design checking purposes, into Classes 0, 1 2 and 3 – see Table 9.2. The requirements of Category 3 checks are significant in time and cost in procuring and paying for an independent design checker.

The categories align to the different risk levels of the work, examples of which are shown in Table 9.2.

Class 0 includes basic construction methods which are low risk; Class 1 are routine construction methods that are low to medium risk; Class 2 includes specialist construction methods (medium to high risk) and Class 3 covers complex, unusual and bespoke construction methods which are high risk. Sometimes, it is not the method that dictates the level of design checking. If the surrounding area has high-risk features or processes, then the contract may stipulate independent design checking. This is the case in sites adjacent to railways, chemical plants and so on.

The design of the temporary works should be based on the agreed design brief. Any proposed alteration or modification of the design brief by the designer should be referred back to the TW coordinator. Temporary works designers include the manufacturers and suppliers of proprietary TW equipment and those working in a contractor's temporary works department or office.

Figure 9.3 is an example of a TW design certificate (Category 0–2), and Fig. 9.4 is a design certificate check which is used for Category 3.

Table 9.2 Categories of design check (BS 5975).

Design complexity risk	**Type of design**	**Independent design checker required?**
DCROO	Manageable design as per industry safe standard (CFRO/CFR1)	No – site/design team member can check
DCRO	Standard solution with minimal consequence of failure (CFRO/CFR1)	N No – site/design team member can check o
DCR1	Simple design with minimal consequence of failure	No – site/design team member can check
DCR2	More complex design with failure risk GFR2 or below	Yes – someone not involved in the design team
DCR3	Complex/innovative designs – not CFR3	Yes – by another organisation
DCR4	Abnormal/innovative design with a CFR3 category	Yes – by another organisation

Temporary Works Design Certificate

No.:

Contract Name and No.	**Designer(s)**	**DB No.**	Structure	Design check category (0,1 2, 3)

AFFIRMATIONS		
Reasonable professional skill and care have been used in the design of the following temporary works:		
TW	*Signed*	*Date*
The following drawings/documents accurately reflect the above		
		Date
To ensure compliance with the Contract, state the codes and standards (and any derivations - including justification for their use) used in the design.		
Additional comments		

Acceptance by TW designer	**Design team leader** *I certify that the staff who has prepared the above design Is competent to carry out their duties and that (so far as I can reasonably ascertain), they have used reasonable professional skill and care*
Name:	Name:
Signed:	Signed:
Date:	Date:

Figure 9.3 An example TW design certificate.

Temporary Works Design Certificate | *No.:*

Contract Name and No.	**Designer(s)**	**Structure**	**DB No.**	**Design check category** (0,1 2, 3)

AFFIRMATIONS		
Reasonable professional skill and care have been used in the design of the following temporary works:		
TW	*Signed*	*Date*
The following drawings/documents accurately reflect the above		
		Date
Description of checks carried out, e.g. Concept, Structural, Dimensional and how they comply with the Contract, codes and standards (and any derivations) used in the design.		
Additional comments		

Acceptance by TW designer	**Design team leader** *I certify that the staff who has prepared the above design check Is competent to carry out their duties and that (so far as I can reasonably ascertain), they have used reasonable professional skill and care*
Name:	Name:
Signed:	Signed:
Date:	Date:

Figure 9.4 An example TW design check certificate.

9.7 Scaffolding

Working scaffolding is defined in the BS EN 12811 Temporary works equipment as 'temporary construction, which is required to provide a safe place of work for the erection, maintenance, repair or demolition of buildings and other structures and for the necessary access'. It can also be used for inspection during the construction process.

There are many different types of scaffolding:

- Supported – this is the most common type and is built upwards from the base.
- Suspended – either from a roof or another tall construction, where building up from a base is not feasible when access to upper levels is required.
- Rolling – instead of a stable base, this type of scaffolding is on (lockable) castors and may be used to access a structure extending over a long distance.
- Mobile – ease of access and the level of movement on the proposed scaffolding are factors to take into consideration when deciding upon mobile or fixed scaffolding.
- Aerial lifts – these are used where workers need to be able to access a number of levels in order to be able to complete a construction.

Scaffolding tied to a building increases stability. It can also be braced laterally using façade and ledger bracing. Scaffold tube is the most common type of material used in scaffolding in the United Kingdom.

It is galvanised and usually comes in two thicknesses: 3.2 or 4 mm. The scaffolding is constructed using clamps with the tube capacities being limited by the safe slip load capacity of the coupler. This is much lower than the actual tensile resistance of the tube. 'Scaffolding is designed for its self-weight, i.e. the weight of the boards, tubes, guardrails, toeboards etc. and imposed loads such as wind. The imposed load applied to the scaffolding depends on its use' (Designing Buildings Wiki, 2015).

HSE guidelines categorise the erection of scaffolding as a specialist task by people with training appropriate to the type and complexity of the scaffolding being erected. Hence, most scaffolding erection is sub-contracted to scaffolding contractors unless there is in-house competence. However, there must be appropriate levels of supervision, depending on the complexity and the experience of the personnel involved.

Steps need to be taken to ensure that the erection, alteration and dismantling of all scaffolding structures (basic or complex) are carried out under the direct supervision of a competent person. For complex structures, he/she would be an Advanced Scaffolder. All operatives would need to be aware of the safety guidance (and the latest changes to it) and good working practices (e.g. NASC's TG20:13). This may be achieved through briefings or toolbox talks.

Scaffold design

'It is a requirement of the Work at Height Regulations 2005 that unless a scaffold is assembled to a generally recognised standard configuration, e.g. NASC Technical Guidance TG20 for tube and fitting scaffolds or similar guidance from manufacturers of system scaffolds, the scaffold should be designed by bespoke calculation, by a competent person, to ensure it will have adequate strength, rigidity and stability while it is erected, used and dismantled' (HSE, 2015a). Certain information needs to be provided to the scaffolding contractor to ensure safety and efficiency. This includes:

- Site location
- Period of time the scaffold is to be in place

- Intended use
- Height and length and any critical dimensions which may affect the scaffold
- Number of boarded lifts
- Maximum working loads to be imposed and maximum number of people using the scaffold at any one time
- Type of access onto the scaffold, for example staircase, ladder bay and external ladders
- Whether there is a requirement for sheeting, netting or brickguards
- Any specific requirements or provisions (e.g. pedestrian walkway, restriction on tie locations and inclusion/provision for mechanical handling plant, e.g. hoist) of the ground conditions or supporting structure
- Information on the structure/building the scaffold will be erected against together with any relevant dimensions and drawings
- Any restrictions that may affect the erection, alteration or dismantling process.

Using this information, the scaffolding contractor can provide the following information (HSE, 2015a):

- Type of scaffold required
- Maximum bay lengths
- Maximum lift heights
- Platform boarding arrangement safe working load/load class
- Maximum leg loads
- Maximum tie spacing both horizontal and vertical and tie duty
- Details of additional elements such as beamed bridges, fans and loading bays, which may be a standard configuration or specifically designed
- Information can be included in relevant drawings, if appropriate
- Any other information relevant to the design, installation or use of the scaffold
- Reference number, date and so on to enable recording, referencing and checking.

The loading relating to scaffolding comes under four classes (TG20:13):
Service Class 1 - 0.75 kN/m² - inspection and very light duty access
Service Class 2 - 1.5 kN/ m² - Light duty such as painting and cleaning
Service Class 3 - 2.0 kN/m² - General building work, brickwork and so on.
Service Class 4 - 3.0 kN/m² - Heavy duty such as masonry and heavy cladding

Table 9.3 shows the scaffold structures that need a bespoke design.

Loading

Wind loads shall be calculated by assuming that there is a velocity pressure on a reference area of the working scaffold, which is in general the projected area in the wind direction. The calculation can be done through commercial software that has online facilities to calculate the maximum safe height and the minimum required ties based on the wind conditions at the site location. The wind loads on the scaffolding will depend on the covering/debris nets used. The covering's large surface area increases vulnerability to wind damage, either affecting the covering's integrity or the covering being blown over taking the scaffolding with it. The level of wind load will necessitate checking the ties, their capacities and frequency.

Scaffolding safety

The risk of falls, which account for over half the fatalities in the construction industry, is high for scaffolders. Wherever practicable, general access scaffolding should be used for working at height. Scaffolding should not be overloaded (see *Loading* section) nor insufficient ties used or removed too early (HSE, 2006).

Table 9.3 The scaffold structures needing a bespoke design.

Marine scaffolds	Temporary ramps
Buttressed free-standing scaffolds	Sign board supports
Boiler scaffolds	Rubbish chute
Temporary roofs and temporary buildings	Spectator terraces and seating stands
Power line crossings	Temporary storage on site
Lifting gantries and towers	Radial/splayed scaffolds on contoured facades
Complex loading bays	Sealing end structures (e.g. temporary screens)
Steeple scaffolds	Access scaffolds with >2 working lifts
Free standing scaffolds	Masts, lighting towers and transmission towers
Towers requiring guys or ground anchors	Advertising hoardings/banners
All shoring scaffolds (dead, raking, flying)	Pedestrian footbridges or walkways
Slung and suspended scaffolds	Cantilevered scaffolds
Truss-out scaffold	Façade retention
Pavement gantries	Bridge scaffolds
Mobile and static towers	Elevated roadways

As a minimum requirement, every scaffold gang should contain a competent scaffolder who has received training for the type and complexity of the scaffold to be erected, altered or dismantled. Operatives are classed as 'trainees' until they have completed the approved training and assessment required to be deemed.

The Health and Safety Executive (2006) stipulate that guard rails need to be used to protect workers and materials (e.g. brick guards) stored at height. Foam scaffold tubes at ground level can help protect pedestrians. Several precautions are needed to ensure safety on and around scaffolding:

- Platforms should be fully boarded and wide enough for the work and for access (usually at least 600 mm wide) with properly supported boards and no excessive overhanging (e.g. no more than four times the thickness of the board).
- There should be safe access onto the work platforms, preferably from a staircase or ladder tower.
- Loading bays should be fitted with fall protection, preferably gates, which can be safely moved in and out of position to place materials on the platform.
- The scaffold should be suitable for the task and checked whenever it is altered or adversely affected.

Guard rails, toe boards and brick guards should be strong and rigid enough to prevent people from falling and be able to withstand other loads likely

to be placed on them. These precautions should include:

- Main guard rail at least 950 mm above any edge from which people are liable to fall
- Toe board and brick guards where there is a risk of objects rolling or being kicked off the edge of the platform
- Sufficient number of intermediate guard rails or suitable alternatives positioned so that the unprotected gap does not exceed 470 mm.

Barriers other than guard rails and toe boards can be used, so long as they are at least 950 mm high, secure and provide an equivalent standard of protection against falls and materials rolling or being kicked from any edges.

Scaffold inspection

The responsibility for inspection rests with users and hirers. Inspections should be carried out following installation or before first use and at an interval of no more than every 7 days thereafter. An inspection is required after any event that jeopardises the safety of the scaffold. Inspections should be undertaken by 'a competent person whose combination of knowledge, training and experience is appropriate for the type and complexity of the scaffold' (HSE, 2015a). Their report should cite any defects and the action required to remedy them.

9.8 Falsework

Falsework generally relates to the structural vertical support of concrete decks and so on, whereas formwork is that used to form soffits, walls and columns. The British Standard Code of practice for temporary works procedures and the permissible stress design of falsework (BS5975: 2008+A1:2011) describes the difference between falsework and temporary works:

- Falsework is a temporary structure used to support a permanent structure while it is not self-supporting.
- Temporary works are parts of the works that allow or enable construction of, protect, support or provide access to, the permanent works and which might or might not remain in place at the completion of the works.

There are three main types of systems used for falsework:

- Aluminium support legs with aluminium frames assembled into falsework systems.
- Individual aluminium or steel props, including either timber header beams or proprietary panels.
- Heavier steel falsework.

The design of falsework differs from permanent works in that they are highly stressed over short periods of time. Self-weight keeps it in place, but it needs to be designed to carry all vertical, horizontal and inclined loads imposed upon it by the structure.

Foundations are needed for extensive/complex falsework. This includes construction operations and wind loading. The capacity of falsework can be provided by the manufacturer. The fact that the components of the falsework may be re-used many times needs to be taken into account in the design.

BS EN 12812:2008 (BSI, 2008) specifies performance requirements for the design of falsework in accordance with one of the three classes: A, B1 and B2. Class A covers falsework for simple constructions such as in situ slabs and beams and should only be adopted when:

- Slabs have a cross-sectional area not exceeding 0.3 m^2 per metre width of slab;
- Beams have a cross-sectional area not exceeding 0.5 m^2;
- The clear span of beams and slabs does not exceed 6.0 m and
- The height to the underside of the permanent structure does not exceed 3.5 m.

Class A has dimensional limitations, which will generally restrict its application to building work. No specific structural design rules are given within this class. Class B1 is based directly on the Eurocodes series (EN 1990, ENV 1991 to ENV 1999), with the design process and all documentation being to the standard of permanent works design.

Class B2 is based on a lower level of calculation. It takes second-order effects into account and contains some information on simplified methods.

Falsework. Performance requirements and general design - European Standard BS EN 12812:2008

This European Standard gives performance requirements for specifying and using falsework and gives methods to design falsework to meet those requirements. Clause 9 provides design methods. It also gives simplified design methods for falsework made of tubes and fittings. The information on structural design is supplementary to the relevant Structural Eurocodes.

The lateral stability of falsework should be considered with beam grillages used, which are designed to be able to resist, at each phase of construction, the applied vertical loads (*W*) and a horizontal disturbing force FH which is the greater of:

- 2.5% of the applied vertical loads (i.e. 2.5%W) considered as acting at the points of contact between the vertical loads and the supporting falsework or
- The forces that can result from erection tolerances (normally taken as 1% of the applied vertical load (i.e. 1%W)) plus the sum of other imposed loads, including wind, out of vertical by design, concrete pressures, water and waves, dynamic and impact forces and the forces generated by the permanent works (BSI, 2008).

9.9 Formwork

Formwork is a temporary mould into which concrete is poured and formed. It is usually fabricated using timber, but it can also be constructed from steel, glass fibre reinforced plastics and other materials. Formwork may be part of the falsework construction and is used in four main areas: walls, columns, beams and slabs. When it is used for the underside of suspended slabs and beams it is known as soffit formwork (Concrete Society, 2010).

There are three main types of formwork: engineered systems, timber and re-usable plastic. The choice of type needs to take into consideration the type of concrete and the temperature of the pour.

Steel formwork has the advantage of speed of construction with the formwork built from pre-fabricated modules. Its life-cycle costs are lower as the frame is almost indestructible and can be used thousands of times.

Timber formwork is constructed on-site using timber and plywood materials; whilst easy to use, it is time-consuming for large structures, being very labour intensive. It is a very flexible material that is used in complex sections of formwork.

Plastic formwork is also re-usable, employing interlocking and modular systems. It is lightweight and is used for simple concrete structures and is suited to repetitive designs such as mass housing schemes.

Stay-in-place formwork is assembled on-site using pre-fabricated fibre-reinforced plastic. This type of formwork is often used for concrete columns and piers where it 'stays in place, acting as permanent axial and shear reinforcement for the structural member. It also provides resistance to environmental damage for both the concrete and reinforcing bars' (Designing Buildings Wiki, 2015).

An important factor to be taken into account when selecting formwork is the rate of pouring concrete. Numerous failures have occurred when exceeding the rate of pour for a specific concrete mix design as stipulated by the formwork designer.

The point at which the formwork can be removed (struck) depends on the rate of strength gain of the concrete. It is a balance between speed of construction and safety. This will differ for vertical and horizontal structures. A minimum value of $5\,N/mm^2$ is recommended in all cases when striking vertical formwork so as not to damage the permanent concrete in the process (Designing Buildings Wiki, 2015). Inspection of the final concrete structure is important for strength and aesthetics. Further factors that should be taken into consideration are:

- Colour variation
- Finish
- Honeycombing
- Moisture loss, affecting hydration
- Durability
- Permissible deflection
- Frost damage
- Further mechanical damage due to site operations
- Mechanical damage due to the removal of formwork
- Thermal cracking and shock.

9.10 Earthworks

Earthworks are defined as (BSI, 2009):

- Structures formed by the excavating, raising or sloping of ground, for example embankments, cuttings or remediated natural slopes.
- Civil engineering process that includes extraction, loading, transport, transformation/improvement, placement and compaction of natural materials (soils and rocks) and/or secondary or recycled materials, in order to obtain stable and durable cuttings, embankments or engineered fills.

Table 9.4 Soil properties – bulking and shrinking.

Material	Bulk density (Mg/m³)	Bulking factor	Shrinkage factor	Diggability
Clay (low PI)	1.65	1.30	—	M
Clay (high PI)	2.10	1.40	0.90	M-H
Clay and gravel	1.80	1.35	—	M-H
Sand	2.00	1.05	0.89	E
Sand and gravel	1.95	1.15	—	E
Gravel	2.10	1.05	0.97	E
Chalk	1.85	1.50	0.97	E
Shales	2.35	1.50	1.33	M-H
Limestone	2.60	1.63	1.36	M-H
Sandstone (porous)	2.50	1.60	—	M
Sandstone (cemented)	2.65	1.61	1.34	M-H
Basalt	2.95	1.64	1.36	H
Granite	2.41	1.72	1.33	H

For estimating purposes, earthworks include:

- Excavation
- Grading: moving earth to change elevation
- Temporary shoring
- Back fill or fill: adding earth to raise grade
- Compaction: increasing density
- Disposal.

Price considerations include:

- Material type (diggability)
- Water level and moisture content
- Length of haul
- Haul road condition.

Estimating the amount of material involved should allow for the bulking up of excavated material – see Table 9.4. There are online services which calculate the level of bulking and the skip/lorry size required for disposal/relocation.

Earthmoving equipment (see ***Plant*** honeycomb for more details) includes:

- Bulldozer
- Drag line
- Dump truck
- Shovels
- Hydraulic excavators
- Grader
- Rollers.

Excavation

Excavation includes almost any operation involving the ground, for example:

- Excavations for foundations
- Site clearance
- Investigation works
- Archaeological digs
- Digging new drainage ditches
- Trenching operations for laying new services
- Contaminated land removal
- Post holing for lighting and fences.

Excavations need to be kept free of water, which may involve pumping or drains. Therefore, it is important to have information on the level of the water table and the soil strata/type. Any pumped water needs to be disposed of without affecting the stability of other parts of the site/surrounding area (BSI, 1989).

Cofferdams can be part of temporary works, providing a dry work area in a water environment, and dismantled once the permanent works are in place. CDM (2015, p. 52) states that cofferdams should be of 'suitable design and construction and appropriately equipped so that workers can gain shelter or escape if water or materials enter it; and be properly maintained' (HSE 2015a).

A trench is an excavation where the length is greater than the depth. They are usually excavated to form foundations or to take services such as pipes and cables. They can be dug out by hand or machine. Trench support is very important for the safety of workers and the ability to dig deeply. This can be made out of timber, particularly in low-risk situations or narrow trenches/shafts. This is labour-intensive. There is no minimum safe depth for trenches as all excavations are to be treated on merit. All side supports must be designed by a competent engineer. Not only has soil loads to be taken into account but also forces exerted by adjacent structures, plant movements and so on.

The earthworks team should aim to provide an earthworks design that is feasible, functional, constructible and suitable for the proposed end use. Consideration should be given to land requirements, including all temporary works. The design should be developed to minimize environmental impact during the construction phase, in use and for future maintenance operations.

BSI (2009, p. 13)

Trench boxes can be used for support and are placed in pre-excavated trenches or a 'dig and push' technique (Designing Buildings Wiki, 2015). Trench sheeting, with extensive strut support, can be used to retain soil, but loading is an important consideration due to the thickness of the sheets.

Regulation 22 of the CDM regulations (HSE, 2015b) set out the safety requirements for excavations. Supports and battering of excavations are essential requirements to maintain a safe working environment. Care should be taken about the area surrounding the excavation to ensure no people, material or equipment can fall into the excavation and that no part of an excavation or ground adjacent to it from being overloaded by work equipment or material. Regular inspections are an important part of the safety regime for excavations.

Dewatering

Dewatering is the control of excessive groundwater by pumping or other methods. Controlling surface water can be achieved by intercepting/diverting/collecting run-off and reducing the amount of water generated on-site. Drains, tanks and flood alleviation are all ways of dealing with excess water on-site.

Ground freezing is a process of making water-bearing strata temporarily impermeable and to increase their compressive and shear strength by transforming joint water into ice. It can be used for groundwater cut-off, earth support, temporary underpinning, stabilisation of earth for tunnel excavation, to arrest landslides and to stabilise abandoned mineshafts.

To freeze the ground, a row of freeze pipes are placed vertically in the soil. Care needs to be taken when the soil thaws, as its water content will have expanded in the freezing process.

9.11 Temporary services

Figure 9.5 A generator providing a temporary electrical supply.

The production process needs energy, water and communication facilities to operate. The utilities will serve the permanent facility, but production on-site requires temporary services to be installed. Their installation costs are part of the estimate as well as the associated temporary infrastructure, operation and maintenance.

A temporary supply usually requires a secure and watertight cabinet to be installed within the site boundary containing meter tails (cables), consumer unit and power point already installed by a qualified electrician.

Where the installation of a temporary electrical supply is not possible, generators can be used (see Fig. 9.5). They can be used to power large items of plant, for example tower cranes where there is insufficient mains power available. The size of the generator needed can be calculated by identifying the tools that will be powered by the generator, total the wattage they require if used simultaneously and then select the generator that has continuous rated watts that meet or exceed this total. Examples of wattage for selected tools/equipment are shown below:

Circular Saw 1200w
9" Angle Grinder 2350w
Cement Mixer 1320w
10-16" Chain Saw 1500w
Arc Welder 3500w

Temporary water connections

Costs for a new water connection can vary greatly and are dependent on a range of factors, such as where the main is in relation to the site entry point and whether the supply company has to excavate unmade ground, pavements or roads to make the connection. There are two main components to the connection charges. The first of these is the connection charge itself and the second is an infrastructure charge which is levied to take account of the additional load of a new user on the entire water supply and waste water systems. The infrastructure charge is the same for water and waste water. If the site is a long way from a public main and it becomes necessary for the water company to extend the main, these costs will be passed on to the contractor although an adjustment will be made to reflect the value added by future revenue generation from the additional main and connection. Information required for application for temporary water supply and drainage:

Extract from **BS 7671 'Requirements for electrical installations'** *BS 7671 requires that every installation shall be inspected and tested to verify that the Regulations have been met before being put into service. The requirements are stated in the following Regulations: 133-02-01 On completion of an installation or an addition or alteration to an installation, appropriate inspection and testing shall be carried out to verify so far as is reasonably practicable that the requirements of this standard have been met. 711-01-01 Every installation shall, during erection and on completion before being put into service be inspected and tested to verify, so far as is reasonably practicable, that the requirements of the Regulations have been met. Precautions shall be taken to avoid danger to persons and to avoid damage to property and installed equipment during inspection and testing.*

- Detailed site layout plan for temporary/site investigation works showing:

1. Site access points
2. Designated service strip
3. Existing utility information
4. Extent of developer's ownership
5. Area for adoption by the local authority
6. Wayleaves, private land information

- Location plan showing nearest public highway and surrounding geography
- Details of any rainwater harvesting or grey-water systems to be incorporated
- Temporary buildings and location of temporary water connection required
- A land risk assessment if there is a chance that the new supply pipes will cross any potentially contaminated land
- Soil survey
- Permits and authorisation requirements for the site
- Health risks
- Safety hazards
- Type of connection required

1. Boundary box
2. Wall box
3. Internal (flats and apartments only)

- Fire supplies – hydrants fire hose reels and so on
- Details of sprinkler systems if applicable
- Supply dimensions
- Maximum daily demand
- Maximum instantaneous demand
- Volume of on-site water storage
- Phasing
- Occupancy rate
- Drainage – surface water, connection to main sewer and soakaways plan required

9.12 Façade retention

Although façade retention involves some form of scaffolding/support process, it is shown separately here because of its significant cost to a project and other factors involved, for example, protection measures and listed building consent.

A design may require that the original façade is kept so that the new building is in keeping with the surrounding area – mostly used for listed buildings. Whilst the façade retains historical significance, the interior can be constructed and fitted out to meet modern needs. Shoring is needed to stabilise and maintain the façade in its original position, which is then removed once the new build is tied in.

The spread of the base of the shores will depend on site layout and the surrounding highways and footpaths. Consideration should be made of adjacent/integral cellars and basements and the position of services infrastructure. 'The greater the spread of the base of the shores the lower the loads in the structure and the foundations' HSE (1992)

There are a number of retention types:

- Scaffolding, suitable for low level facades between three and four storeys, with sufficient space at their base for installation.
- Proprietary retention, involving props, ties and bracing suitable for higher facades as the general quantity of components is reduced.
- Fabricated steelwork, used when the cost of hiring proprietary equipment over long periods of time outweighs the cost of fabricating a structure
- Combinations of fabricated and proprietary retention systems.

To avoid cracking the façade, excessive movement must be eliminated by using substantial support systems which reduce/eliminate movement in any direction. Wind loads and impact loads must be taken into account. Once the façade is considered to be self-supporting (with the appropriate shoring), then wind is the biggest threat to its stability. *BS EN 1991-1-4:2005+A1:2010Eurocode1: Actions on Structures* gives more information and details about wind load calculations.

There are a number of considerations in the development of façade retention:

- The timing of both full vacant possession and possible access for prior investigations
- Physical access for investigations of the building and relevant parts of neighbouring properties and by whom this will be done
- The necessary consents for the construction and any diversions of services required or road/footpath closures
- The extent of demolitions required and any repairs needed before this or before construction begins
- Details of fixings, bracing, protection and any consolidation to historic facades that need to be agreed with the client and/or conservation architect and local authority officer
- Whether or not underpinning is required
- Party wall issues
- Constraints imposed by permanent works.

9.13 Structures – design and loading

Any contractor-installed temporary works must be checked to ensure that the following loads are taken into account:

- Construction and equipment loads
- Dead loads (i.e. concrete pours and erected beams)
- Live loads
- Wind loads
- Lateral earth pressure loads
- Jacking loads (if jacks bear on temporary works)
- Any other loading on the temporary works as a result of construction activities.

9.14 Plant foundations

Plant foundations for say, cranes and crane outriggers, are two examples of temporary foundations. The main differences between temporary and permanent foundations are that:

1. The construction of the temporary foundations is often on the surface of the ground
2. Temporary foundations are likely to be subjected to their full design load, whereas a permanent foundation might never be subjected to its full design load.

There are different types of crane foundations according to the size and type of crane, the ground conditions and the location constraints.

Depending on the ground conditions, the foundations may be a compacted structural fill, concrete foundations or in poor soil conditions, piled foundations.

The designer of the foundation should produce a report on the criteria and calculations used for the design, including the values for maximum design vertical loads, horizontal loads, applied overturning moments and rotational torques and the maximum out-of-service wind speeds considered.

Foundations distribute the high concentrated loads of the cranes. For mobile cranes, mats are used. These may be made of heavy timbers or fabricated steel mats placed on compacted fill. Mats are placed under outriggers to reduce single point pressure. The mat dimensions will depend on the type of soil/ground conditions.

9.15 Protection on-site

Weather protection

Weather protection may be used to ensure that work can continue even if the weather is inclement. In cold/wet climates, this sort of protection is commonplace. Temporary roofing or temporary shelters can be used, but the wind load needs to be taken into consideration – see BS 8410 'Code of Practice' for lightweight temporary cladding for weather protection and containment on construction works.

There may also be the need to protect some processes, such as frost protection for foundations and mortar.

Protection of archaeological work

Planning Policy Statement (PPS) 5 – Planning for the Historic Environment (replaces PPG16) states that: 'Where an application site includes, or is considered to have the potential to include, heritage assets with archaeological interest, local planning authorities should require developers to submit an appropriate desk-based assessment and, where desk-based research is insufficient to properly assess the interest, a field evaluation'. Therefore, the onus is placed on the developer. The cost of any delay, or works needed to protect any archaeological finds, can be significant, and so the use of an archaeological consultant to appraise the site is important before the developer makes any serious commitment to the project. An archaeological appraisal is a combination of a desk study and site investigations. A field evaluation can be undertaken at the same time as, or in conjunction with, the ground investigations for the construction project. The archaeological costs involved in, say, a test pit evaluation include:

- Breaking out of concrete
- Labour for spoil removal and other purposes

- Direct archaeological costs
- Shoring, barriers and hoists and the like for deeper pits
- Backfilling.

There will also be fees from the relevant consultants involved such as engineers, architects and archaeological consultants.

9.16 Traffic management

Controlling the movement of traffic and pedestrians on-site is imperative for both safety and efficiency. The safety aspect is dealt with in more depth in the CDM honeycomb, while the associated noise, dust and so on are described in the Environmental management honeycomb. Traffic movements involve the delivery, relocation/storage of scaffold, materials and plant. They need to be controlled and monitored to reduce waiting times both on-site and on the adjacent highway. Estimating for traffic management requires an allocation of costs for personnel, equipment and signage.

Traffic routes through the site need to be well signed, with a speed limit enforced; the use of speed ramps may be necessary. Hazards, restrictions and directions should be clearly identified and communicated by signs, signals and instructions. Hazard warning signs should comply with the Health & Safety (Signs and Signals) Regulations 1996. Traffic lights can be used to control flow at busy junctions, in narrow locations and at entry and exit locations to the site. Banksmen (traffic marshals) are trained to direct traffic and bank vehicles safely about the site. A Banksman trained in lifting operations is a separate specialist qualification and not generally held by the standard traffic marshal. The term Banksmen is now generally used as the operator trained in lifting operations and would normally also hold a slinger qualification, that is slinging the loads using chains or strops for the cranes to lift. They wear distinctive 'hi viz' clothing for identification and use an agreed set of standard signals.

9.17 Temporary access for vehicles and pedestrians

Access routes are required on-site for pedestrians, cyclists and road traffic and for the transport of equipment and materials, for emergency vehicle access and maintenance purposes. Where possible, temporary roads should be designed to avoid the need to reverse, for example in a loop. Specifications for turning areas and bends (for larger vehicles) are provided by the UK Freight Transport Association publication 'Designing for Deliveries' (1999) (FTA 2016).

As construction projects become more complex, bigger and higher, the machinery and transport vehicles used are increasingly heavier. Not only the weight but the height, length and width of vehicles likely to be on-site also need to be considered in constructing temporary access. For example, a standard container on a suitable flatbed vehicle is 4.2 m high. Pipebridges, overhead gantries and so on should be clearly identified and height restrictions clearly marked. Lighting may be required for parts of the access road (e.g. for early/late deliveries). Adequate drainage is needed for access roads to avoid flooding. Signage is essential to maintain traffic flows and to ensure safety.

9.18 Barriers for pedestrians and vehicles

Barriers are an effective way of separating pedestrians and vehicles both at entrances and exits to the site and at crossing points within the site. Different types include:

- Concrete
- Metal

- High-density polythene
- Water-filled.

The numbers, dimensions and means of securing the barriers need to be taken into account. Protection may also need to be provided for installations, against vehicle damage. *BS 7669-3:1994 'Vehicle restraint systems'* gives guidance on the appropriate measures. Safety fences can be made of different types of beam: tensioned corrugated; untensioned corrugated; open box beam; rectangular hollow section or a wire rope safety fence. All barriers will need to be maintained and inspected at intervals.

9.19 Haul roads

Haul roads need to be designed to serve a high volume of heavy vehicles. The UK Design Manual for Roads and Bridges (Department of Transport, 2008), the design guidance for temporary haul roads, assumes that all traffic is the heaviest class of HGVs, with an estimated peak of 100 two-way HGV movements per day operating for 365 days a year with a road having a design life of 5 years. This level of traffic would need, as a minimum, a 150-mm thick sub-base of hydraulically bound material (crushed rock coarse aggregate), topped with 100 mm of asphalt (dense bitumen macadam). Height restrictions, with their associated signage, will need to be considered in the planning of haul roads.

9.20 Works associated with any ordnance uncovered

For any major infrastructure project, the nominated undertaker's contractors will carry out a risk assessment for the possibility of unexploded ordnance being found within construction areas. An emergency response procedure will be prepared and implemented by the contractors to respond to the discovery of unexploded ordnance, including notifications to the relevant local authorities and emergency services. CIRIA C681 (2009) gives guidance on unexploded ordnance.

9.21 Signage

'The Regulations require employers to ensure that safety signs are provided (or are in place) and maintained in circumstances where there is a significant risk to health and safety that has not been removed or controlled by other methods. This is only appropriate where use of a sign can further reduce the risk. The other methods may include engineering controls or safe systems of work and may be required under other relevant legislation. Safety signs are not a substitute for those other methods of control' (HSE, 2015c). Signs not only serve to inform and warn those using the site but also to provide public information, about the project, the client or safety concerns.

9.22 Protection of adjacent properties

Depending on the proximity, protection can include shoring to prevent collapse during construction of basement and so on and also temporary protection if party walls are exposed.

Piling

Cofferdams, tunnelling and so on. See 'Plant foundations'.

10 Cost-estimating techniques

The New Code of Estimating Practice (CoEP) looks at the production of an estimate and tender where a detailed design and a bill of quantities is available, and the constituents of rates (labour, plant, materials and specialty contracts) are priced individually and summarised separately for the review meeting and handover. With the many different procurement approaches now being used, there is often the need for the contractor to be involved at the early design stage and for their input into the estimating process to produce budget estimates, order of cost estimates and cost plans.

At an early stage in a project's development, order-of-estimating techniques are employed in order to set budgets and assess the feasibility of a scheme and, where necessary, gain funding. During design development, a cost plan is an important way of producing a design within the client's budget.

The degree of certainty, that is the level of detail available, will dictate which method can be used – see Fig. 10.1. Order-of-estimating techniques depend on historical cost data being available from previous similar schemes.

There are four main estimating methods:

1. Single-rate approximate estimating
2. Multiple-rate estimating using elemental cost plans
3. Analytical estimating
4. Operational estimating.

Single-rate methods

These include superficial (floor area) and building volume. The floor area method is simple, using concepts widely understood and used extensively. It is used for most building projects. The building volume method is used in some European countries, which takes account of storey heights.

Superficial (floor area) method: This method uses historical data from earlier comparable schemes in terms of the cost per square metre of floor area to take account of the project complexity and type of work. Judgement, knowledge and experience are required to select the most appropriate unit price rate. This is a popular method readily understood as relatively few rules are needed to apply this technique. The floor area of a building is defined as that measured at each floor level using the

New Code of Estimating Practice, First Edition. The Chartered Institute of Building.

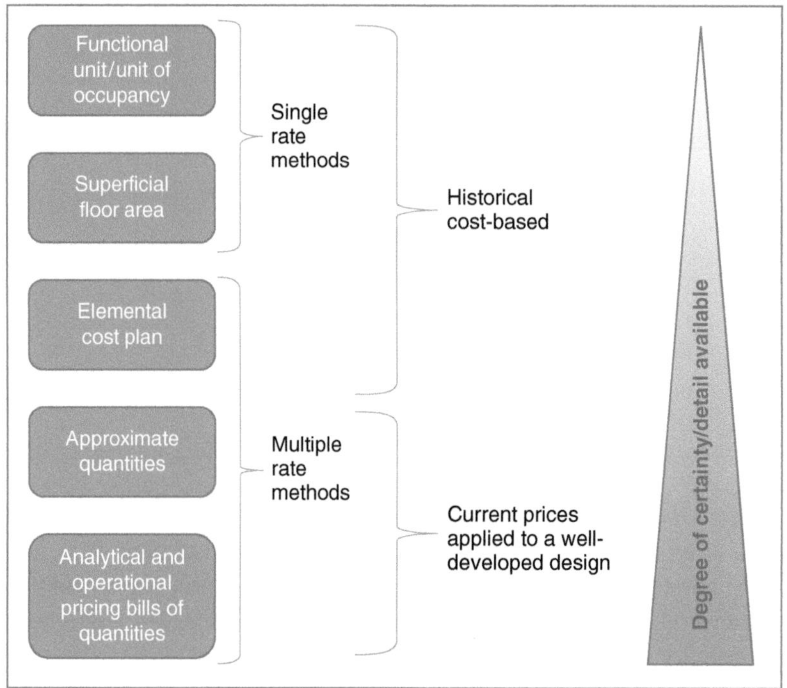

Figure 10.1 The different methods of estimating.

internal dimensions, without making deductions for internal walls or stairs. The gross internal superficial floor area is calculated making due allowance for plant rooms and ancillary items.

As with any estimating technique, a number of adjustments are needed to take account of location, specification, degree of complexity, size, shape, ground conditions and number of storeys. In order to assess these factors, reliable historical costs are needed from a variety of buildings within each building category. A separate assessment must be made of external works, main services and drainage which can all vary substantially, depending on the nature and location of the site.

Multiple-rate methods

Elemental cost plan: This can be produced from a preliminary building design. It is based upon the functional elements of a building, for example substructure, frame and upper floors. The method depends on reliable data being available from comparable projects where the actual costs for each building element are known. Figure 10.2 shows how prices from a bill of quantities for a similar office building have been set against a standard list of elements. In this example, the floor area of the previous building was 2,900 m^2. It is shown that the breakdown of costs for a 3,850-m^2 building can be calculated simply by applying the earlier proportional cost in each element to the second building. A number of adjustments are easily made, such as the introduction of a lift and a piled foundation.

A detailed cost plan is prepared as the design develops through the design stage. Cost checks are regularly made during the design phase to check the element unit price against the budget for the element. Cost planning has evolved to become a popular method of obtaining a bid price from the contractor at the tender stage. It places more emphasis on the work of the estimator who must measure the detailed quantities in order to price the work.

The main failing of the cost planning system at the tender stage is that the cost planning elements do not separate the items into location or sequence, nor do they necessarily reflect the works packages that the contractor will use to build up the price.

PROJECT: Helix Electrics 2-storey offices		KB Electronics Gross floor area (m²) 2,900 Elemental costs			Helix Electrics Gross floor area (m²) 3,850 Elemental costs		
Ref	Elements	£/m²	Totals	Notes	£/m²	Totals	Notes
1	Substructure	91.94	266,625	RC pads	128.00	492,800	Piled founds
2.1	Frame	90.19	261,563		110.20	424,270	
2.2	Upper floors	48.88	141,750		51.20	197,120	
2.3	Roof	68.08	197,438		66.40	255,640	
2.4	Stairs	19.20	55,688		18.70	71,995	
2.5	External walls	84.38	244,688		88.00	338,800	
2.6	Windows and external doors	62.84	182,250		67.50	259,875	
2.7	Internal walls and partitions	41.90	121,250		47.55	183,068	
2.8	Internal doors	8.90	25,819		10.00	38,500	
3.1	Wall finishes	18.04	52,313		21.00	80,850	
3.2	Floor finishes	82.63	239,625		76.00	292,600	
3.3	Ceiling finishes	23.86	69,188		20.75	79,888	
4	Fittings and furniture	12.80	37,125		16.00	61,600	
5.1	Sanitary appliances	7.56	21,938		9.50	36,575	
5.2	Disposal installations	5.94	17,213		7.90	30,415	
5..	Water installations	8.15	23,625		10.00	38,500	
5.4	Heat source and space heating	70.41	204,188		70.75	272,388	
5.5	Ventilation and cooling	27.93	81,000		31.00	119,350	
5.6	Electrical installation	139.66	405,000		140.00	539,000	
5.7	Lift installation	–	0		31.00	119,350	Two lifts
5.8	Security alarms	36.83	106,819		40.55	156,118	
5.9	Fire alarms	14.49	42,019		16.00	61,600	
5.1	Builder's work in connection	9.48	27,506		8.50	32,725	
	Net building cost	974.09	2,824,875		1,086.50	4,183,025	
6.1	Site works	142.27	412,594		149.00	573,650	
6.2	Drainage	32.64	94,669		35.00	134,750	
6.3	External services	9.66	28,013		12.00	46,200	

Figure 10.2 An example of cost plan estimating.

Ref	Elements	£/m²	Totals	Notes	£/m²	Totals	Notes
	Net trade total	1,158.67	3,360,150		1,226.00	4,720,100	
	Preliminaries	89.80	260,415		129.87	500,000	
	Design fees	60.52	175,500			234,850	
	Statutory fees	14.90	43,200			57,750	
	Pre-start costs	–	–		–	–	
	Inflation	–	–			462,000	
	Overheads and profit	74.48	216,000			284,900	
	Contingencies	46.55	135,000			180,950	
	Budget total	1,444.92	4,190,265		1,633.00	6,287,050	

Figure 10.2 (*Continued*)

Accuracy, reliability and confidence are required when using price rates for completed projects that have been published. Tender prices may not reflect the final account price. There is no substitute for building up prices using information obtained from experience and from specialty contractors and materials and component suppliers. Markets are dynamic, constantly changing and reflecting changes in market conditions, particularly labour rates.

The RIBA Plan of Work 2013 provides a process map of the sequence of the design process with gateways that are passed as the design moves from Stage 0, strategic definition, through the design process to Stage 4, technical design (the stage before a tender is sought).

The RIBA Plan of Work 2013 organises the process of briefing, designing, constructing, maintaining, operating and using building projects into a number of key stages. It has been a bedrock document for the architects' profession and the construction industry, providing a shared framework for the organisation and management of building projects that is widely used as both a process map and a management tool and providing important work stage reference points used in a multitude of contractual and appointment documents.

Analytical/composite unit rate estimating

A large part of an estimator's time is devoted to calculating unit rates for items in a bill of quantities. In addition to analytical unit rate pricing, the estimator will use the following techniques:

- Spot items
- Operational estimating.

Spot items: These are operations which are difficult to break down into discrete items of work in a bill of quantities. For example, the demolition of small buildings or the formation of openings through walls is priced by looking at the extent of the work during a site visit. For estimating purposes, spot items may be treated in several ways:

- Approximate quantities can be taken off and unit rates used to calculate a lump sum estimate for the item.

- The description within the bills can be analysed into its constituent operations and trades and an estimate of the cost made for each.
- When the description within the bill is analysed into constituent operations and found to have a predominant trade, then a gang or operational assessment can be made on a time, plant and material basis so that the overall cost can be calculated.

The cost of labour, plant and materials should be separated in accordance with the general principle already described for unit rates (See **Pricing the works**). The work must be inspected thoroughly at the site visit and, where necessary, construction method must be established and documented.

Adequate allowances must be made for storage, temporary works, including supports, access, double handling, small deliveries, making good and reinstatement. Transport can be included in the item but will usually be included in project overheads. If bill descriptions are not clear, or if further information or measurements are required, it may be necessary to revisit the site.

Operational estimating

This system is adopted when the estimator needs to consider the overall duration of an operation and its interrelationships with other trades. This is the case with civil engineering construction or the earthworks and concrete elements of a building project. In these cases, it is unrealistic to look at a single unit of work and wrong to assume that the total cost of an operation is the product of the unit rate and quantity.

For example, a contractor may make an assessment for laying precast concrete manhole rings on the basis of the number a drainage gang can fix in 1 day. For a 2.10-m diameter manhole and 600-mm high units, the estimator might assume that 15 units can be handled, lifted into position and securely bedded in 1 day. If a project has 25 precast concrete units, an allowance of 2 days may be needed because it might be difficult to deploy the plant elsewhere for a small part of the second day. It is unlikely that a simple unit rating exercise would have included an allowance in this way for standing time.

Operational estimating depends on a careful study of how a section of work will be carried out in practice. It is difficult, for example, to price the fixing of roof trusses without looking at working methods.

Case study 1

A new dental surgery has a rectangular plan shape, 60-m long, with 55 nr timber roof trusses above first floor level spanning 8.50 m between wall plate supports. The estimator has drawn up a list of resources for fixing trusses as follows:

Mobile crane (1 nr)	3 days @ 380.00	= 1,140.00
Banksman (1 nr)	24 h @ 10.00	= 240.00
Carpenters (2 nr)	80 h @ 18.00	= 1,440.00
Supervision	10 h @ 20.00	= 200.00
Nails, screws and fixings for 55 trusses		150.00
		£3,170.00
Plus 10% allowance for inclement weather and need for adaptation of truss		£317.00
Cost per truss for labour and plant		**£63.40**

The cost of the trusses delivered to the site will be added with an allowance for waste and any double handling of the materials. In this case, a method statement was not produced. But for more complex construction operations, for example, more planning is needed together with method statements.

Case study 2

A large distribution warehouse has a floor slab with 600 m^3 of concrete to be cast in continuous pours with all joints and reinforcement to be introduced during the casting operation. For the method chosen, 200 m^3 of concrete will be placed each day. After a detailed review of the design and discussions with the consultants, it was agreed to replace steel reinforcement with self-compacting fibre-reinforced concrete that will be pumped using ready mix concrete delivered to the site. Labour rate includes cleaning tools and equipment of 0.50 h/day. Shuttering measured separately, materials measured separately with 12.5% allowance for waste and waste removal.

The estimator and planning engineer drew up a list of resources and durations:

Trade foreman (1 nr)	3 days × 9 h @ £14.00	378.00
Gangers (2 nr)	3 days × 9 h @ £12.00	648.00
Operatives (10 nr)	3 days × 9 h @ £11.00	2,970.00
Carpenters (2 nr)	3 days × 9 h @ £15.00	810.00
Laser screeder (2 nr)	2.5 days × 9 h @ £14.00	756.00
Driver (1 nr)	3 days × 9 h @ £15.00	405.00
Dumper (2 nr)	3 days × 9 h @ £5.00	270.00
Drivers (2 nr)	3 days × 9 h @ £14.00	756.00
Power floats (2 nr)	3 days × 9 h @ £5.00	270.00
Temporary protection during curing		150.00
Total cost for 600 m^3 for labour and plant		**£7,413.00**

In order to insert a unit rate in the bill of quantities, the total cost for placing concrete is divided by the quantity.

Unit rate for labour and plant = £7,413/600 m^3 = £12.35/m^3.

10.1 Approximate quantities

Approximate quantities are needed where the other approximate estimating techniques do not produce sufficient information for a reliable budget. For example:

- A shorter bill of quantities (BoQ) with composite items. An item for external walls, for example, would include both skins of masonry, forming the cavity, wall ties, plastering and pointing. In this case an approximate BoQ is produced and priced with rates taken from a number of sources including previous bills of quantities, price books or guide prices from specialist trade contractors and suppliers. The accuracy of this method will depend on the extent to which the design has been developed. If approximate BoQs are to be used for cost planning during the design stage, they should follow an elemental bill format giving estimated costs for each building element.
- A contractor's BoQ produced from drawings and specifications that includes fewer ancillary items than are required by the standard method of measurement. The rules for measurement, such as deducting openings in walls, are often ignored, as it is assumed that the over-measure will account for the extra labour and increased waste on materials. Computer-aided estimating systems

provide a quick method for creating a bill of quantities. The estimator selects items from a library of descriptions which were previously priced. Resource costs can be changed when material quotations are received.

- In order to start a contract early, a BoQ (often from another project) can be used to establish a tender price. The JCT Standard Building Contract with approximate quantities is the variant for this arrangement.

Case study 3

A developer builds and operates accommodation for students in many towns and cities in the United Kingdom. In order to ensure a strong business case, a unit cost per study bedroom was established. The value of a student bedroom often depends on local demand, land values and finance costs. This case study is for a 200 bedrooms development in Southampton. The business plan has been optimised to produce a reasonable return on the investment and indicates an affordability construction cost of £16,000,000. A construction company, who worked closely with the developer, was asked to produce an initial cost plan for the scheme to test the viability of the project and later to produce cost studies to enable the design team to design within the budget.

The contractor's strategy was as follows:

1. Convert the number of bedrooms into a net building area (the sum of all the rooms). Add space for corridors, staircases, lift lobby, and common areas to arrive at a gross internal floor area.
2. Visit the site to gain an understanding of site constraints and ground conditions. These abnormal works items often have a significant bearing on the practicality of the development. On the other hand, some abnormal site costs can be used to negotiate a better (cheaper) land deal.
3. Obtain standard building costs for the superstructure works. These square metre rates are drawn from an analysis of a similar project in a database of building costs. Contractors are in a unique position to gather current trade package costs from projects on-site. Package costs are then converted to element costs in the form of a standard elemental cost plan.
4. Agree on-cost items with the client, such as profit margin and design fees. In this example, these costs were agreed earlier as part of a framework agreement.
5. Produce an outline cost plan, Stage A, with standard costs, abnormals and on-costs. An example is given in Fig. 10.4.
6. If the total construction cost is affordable, then an elemental cost plan will be prepared using data from a similar scheme.
7. Prepare a design-to-cost report, comprising the elemental cost plan and clear statements of what the elemental costs represent in terms of principal quantities and components - see Figure 10.5.
8. Attend design team meetings, monitor the emerging design and suggest ways to remain within the overall target cost.
9. Obtain advice on market prices from specialist sub-contractors.
10. As the scheme design reaches detailed proposals, produce a bill of quantities from scaled drawings and carry out further market testing to confirm the assumptions made in the cost plan.

Using this process, the contractor can provide continuous cost reporting, from the scheme appraisal stage through to agreeing a contract sum.

10.2 Cost planning

Increasingly, contractors are being asked to develop designs and methods which meet affordability targets. Figure 10.3 shows stages in the pre-contract phase of a construction project. The cost plan is developed as a vehicle for continuous reporting and monitoring costs as the design advances.

A simple 'outline' cost plan can be produced at the earliest appraisal stage, and this will be expanded into a 'first-pass' cost plan at Stage 1 'Strategic brief'. An example of a simple cost plan is shown in Fig. 10.4. The standard building costs have been taken from average rates; the on-costs are usually known, which means that the main effort is put into the abnormals – those items which are site-specific or part of an enhanced design. Once the budget has been established with the client, a top-down design-to-cost plan can be produced (see Fig. 10.5) so that the design team understands the cost limit.

On costs are generally fixed by the terms of the framework agreement, and abnormals need to reflect site conditions. The main building costs are then presented in a design-to-cost report which can be in the form of an elemental cost plan or as a list of principal quantities and descriptions. Figure 10.6 is a typical design-to-cost report aimed at designers who need costs translated into simple quantities and statements for a 3- storey school project in Birmingham.

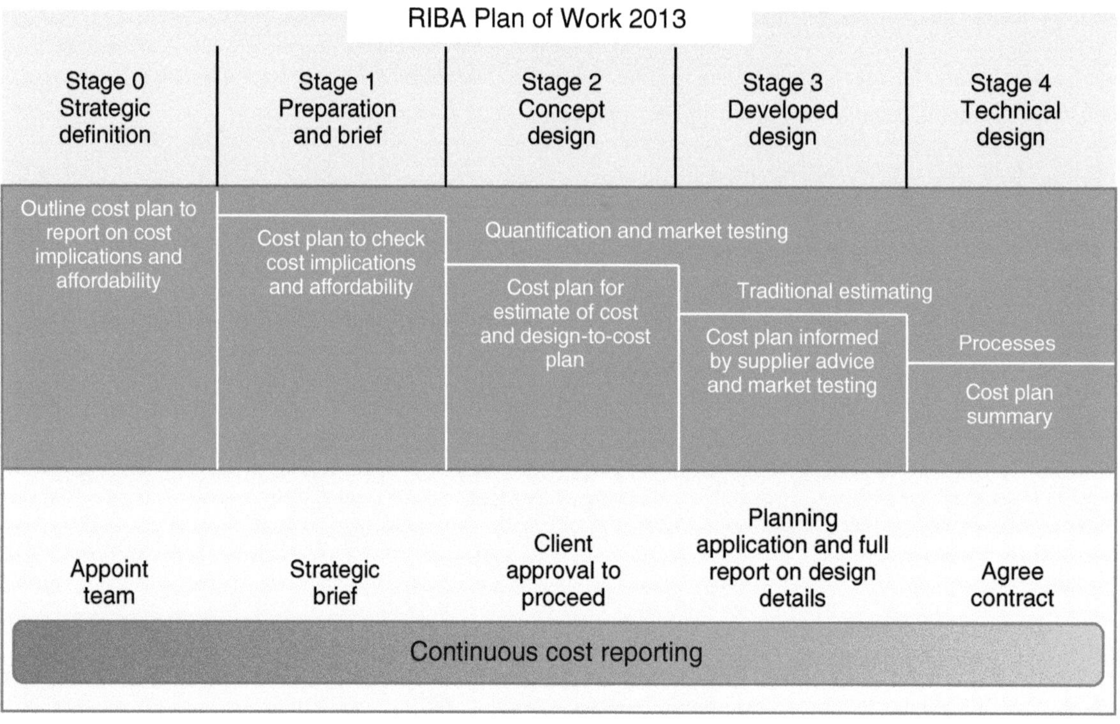

Figure 10.3 Stages in the pre-contract phase of a construction project.

Proposed student accommodation at Leicester				
Overview of capital costs	Schedule areas		£	£/m²
New build-core specification with limited prelims		m²	**12,000,000**	1201.80
Adjustments for design, market factors, and location included				
Building totals	9,985	m²	**12,000,000**	1201.80
Abnormals				
Enabling works - demolition and fencing			**230,000**	23.03
Utilities upgrade for electircal supply			**295,000**	29.54
Waterproofing existing and temporary utilities			**210,000**	21.03
Main sewer diversion			**103,840**	10.40
Buildings and abnormals	net cost		**12,838,840**	1285.81
On-costs		%		
Preliminaries (some items included in elements)		2.00	**256,777**	
Construction risk (included in new build construction price) and contingencies		2.50	**327,390**	
Design fees/statutory charges for approvals		8.00	**1,047,649**	104.92
Pre-construction start costs			**125,000**	16.67
Inflation on build cost to completion		2.00	**256,777**	
Margin on construction work(included in new build construction price)		4.00	**552,191**	
Construction outturn total for design and build			14,852,433	1980.32

Student accommodation - Benchmark data from Exeter student accomodation				
Overview of capital costs		*Areas (sq.m)*	£	£/m²
New build		18,470	**18,561,390**	1004.95
Refurbishment games room		235	**290,507**	1236.20
Building totals		18,705	**18,851,897**	1007.85
Abnormals				
Enabling works - demolition			**212,377**	11.35
Temporary engineering services			**306,388**	16.38
Incoming gas main			**60,492**	3.23
Site works and utilities connections			**185,927**	9.94
Drainage			**192,213**	10.27
Buildings and abnormals	net cost		**19,809,294**	1059.04
		%		
Preliminaries priced separately		13.17	**2,608,885**	139.48
Construction risk priced separately		2.92	**579,079**	30.96
Design fees		8.37	**1,657,474**	88.61
Additional works not anticipated pre-constn		5.08	**1,005,455**	53.75
		included		
		included		
Construction outturn total			25,660,187	1371.84

Figure 10.4 An example of a concept design capital cost plan (m² method).

Top- down Design to Cost Plan Southampton Student Accommodation Block			
No. of bedrooms 200		Gross internal floor area10,000	
	Capital cost affordability		**16,000,000**
On-costs	**Overheads and profit**	–0.055	–880,000
	Sub total		15,120,000
	Labour and materials Inflation allowance in prices	–0.02	–320,000
	Pre-start costs and temporary works	–0.01	–160,000
	Design fees and charges (architectural, engineering, statutory charges)	–0.05	–800,000
	Contingencies/risk	–0.02	–320,000
	Preliminaries	–0.080	–1,280,000
	Abnormals costs		
Abnormals	Enabling works – demolition and fencing		–58,000
	Utilities upgrade (water and electricity)		–20,000
	Site works outside boundary		–100,000
	Drainage – diverting sewer		–350,600
Std costs	New build including design fees, and external works	10,000 m² @ £1600/m²	**16,000,000**
	Refurbishment	-	-
	Target net cost		**16,000,000**

Figure 10.5 A top-down design-to-cost plan.

		Design-to-cost Cost Plan			
		GIFA 7,500 sq m			*Height: 3 storeys*
		Elemental costs			**Notes**
Ref	Element	*Totals*	*£/sq m*	*%*	
1	Substructure	913,795.00	121.84	7.11	Piled foundations, poor ground
2A	Frame	1,136,684.00	151.56	8.85	Structural steel
2B	Upper floors	716,468.00	95.53	5.58	Precast concrete
2C	Roof	274,732.00	36.63	2.14	Steel pitched roof
2D	Stairs and ramps	60,541.00	8.07	0.47	Precast concrete
2E	External walls	1,328,849.00	177.18	10.35	Brick panels
2F	Windows and external doors	257,732.00	34.36	2.01	Timber
2G	Internal walls and partitions	660,643.00	88.09	5.14	Block walls and dry lining
2H	Internal doors	484,100.00	64.55	3.77	Timber
3A	Wall finishes	359,285.00	47.90	2.80	
3B	Floor finishes	251,032.00	33.47	1.95	
3C	Ceiling finishes	293,451.00	39.13	2.28	Suspended ceilings
4A	Fittings, furnishings and equipment	769,275.00	102.57	5.99	Library equipment
5A	Sanitary installations	329,863.00	43.98	2.57	
5B	**Services equipment**				
5C	Disposal installations	175,989.00	23.47	1.37	
5D	Water installations	189,622.00	25.28	1.48	
5E	Heat source	477,494.00	63.67	3.72	Gas heating
5G	Ventilation and cooling	161,397.00	21.52	1.26	No air conditioning, fume cupboard extract
5J	Electrical installations	506,332.00	67.51	3.94	
5J	Lift and conveyor installations	98,000.00	13.07	0.76	Lift for disabled
5K	Communication, security and control systems	174,465.00	23.26	1.36	
5K	Fire alarm system	87,132.00	11.62	0.68	
5N	Builder's work in connection with services	270,796.00	36.11	2.11	
	Net structure cost	**9,977,677.00**			
6A	Site works	185,000.00	24.67	1.44	Fencing, car parking
6B	Drainage	95,600.00	12.75	0.74	Connecting to existing sewer
6C	External services	125,000.00	16.67	0.97	Water and power connections
6D	Enabling works - demolition	75,000.00	10.00	0.58	
	Total including abnormal and ancillary work	**10,458,277.00**			
7	Preliminaries and site based overheads	836,662.16	111.55	6.51	
7	Design fees (Design and build architectural, and structural engineering)	679,788.01	90.64	5.29	
7	Statutory fees	65,000.00	8.67	0.51	
7	Pre-start costs	30,000.00	4.00	0.23	Temporary protection and facilitating works to existing
7	Inflation	156,874.16	20.92	1.22	Inflation allowance for 22 month contract
7	Profit	418,331.08	55.78	3.26	
7	Contingencies	200,000.00	1.36	1.56	
	Budget total	**12,844,932.40**	**1,712.66**	**100.00**	

Figure 10.6 A typical design-to-cost report.

11 Private finance initiative/public–private partnerships/build, operate and transfer, and whole-life costing

There has been a gradual shift away from a focus on lowest capital cost to the consideration of whole-life appraisal, especially with the advent of long-term project investment such as build–operate–transfer (BOT) and public–private partnership (PPP). See the section on ***Procurement*** for a description of private finance initiative (PFI) and other design–build–finance–operate procurement types.

Whole life costing: consideration of the capital and operational/running costs of a facility over its life (service, design, technological, functional, physical).

Whole-life costing takes into account the cradle-to-grave costs – from construction, occupation and operation to disposal. These costs include procurement, maintenance, refurbishment and disposal costs.

The estimator must consider the cost; performance; different user requirements; the running costs for energy, water supply and sanitation; annual and periodic maintenance; insurances; cleaning; security and facility management. The bid team would need to consider how the project can be green and comply with all the changing standards as well as meeting the client's requirements.

Contractors involved in PFI (and similar) projects are having to produce whole-life costs plans as part of their tender bid, to support their planned maintenance requirements for the life of the project, which can be as long as 30 years.

Whole life appraisal: involves consideration of the whole life costs and the performance of the facility over its life.

Data are a major issue in whole-life appraisal. Their collection is costly, complex and changes over time. Whole-life appraisal can fail because of the lack of data and information about the performance and cost of owning and operating facilities in use. This is a major risk in large projects.

Whole life appraisal is the systematic consideration of all relevant costs, revenues and performance associated with the acquisition and ownership of an asset over its physical/economic/functional/service/design life.

In PFI bids and similar contracts, whole-life costs have been developed on a cost-per-square-metre basis using historic data in support. This approach is of little use when trying to choose components or drive down whole-life costs. A better approach is to predict the life of each component in new or existing buildings, establish what maintenance it needs and when it needs to be replaced. The prediction is then costed, thus allowing alternative solutions to be investigated and compared.

The bid team needs to consider a number of areas in a bid for a concession project:

- Understanding the requirements and expectations for a cradle-to-grave performance
- Preparing and/or advising on the schedule of activities

New Code of Estimating Practice, First Edition. The Chartered Institute of Building.

- Preparing early cost plans and cost forecasts (from the feasibility stage)
- Developing target costs plans against the scope and specification and advising on affordability
- Developing and advising on value-based solutions for optimum whole-life cycle. Developing and advising on the financial model for the whole-life cycle.

The PFI process is inevitably long and costly with large multi-professional teams including architects and engineers, accountants and solicitors, along with construction professionals. The tender programme has to be agreed by all parties. The tender would be based on a cost model for the facility which includes:

- Setting up the PFI company/special-purpose vehicle (SPV)
- Land or premises purchase
- Capital funding
- Design fees
- Construction costs
- Operation cost
- Facilities management and maintenance
- Disposal or residual value

Bidding for PFI projects is very long, costly and very specialised. It is beyond the scope of the Code of Estimating Practice (CoEP) to go into the detail of PFI bids. The principles and practice of estimating still apply, but on a much larger scale. Only major companies have the financial resources to take on the risk and cost of bidding for such work.

12 Risk management

12.1 Background

Risk management is seen as good management practice, critical to the survival and management of any business. Risk management is not just a question of top-level analysis. Risk is inherent in every decision, and the risk-aware company requires some assessment of it in every decision made.

The risk management process enables contractors to identify, quantify and manage risks, thereby making better decisions and achieving better outcomes.

Every organisation manages its risk but not always in a way that is visible, repeatable and consistent in how it supports decision-making. The Office of Government Commerce's (OGC) Management of Risk (MoR) system provides an organisation with systems to make cost-effective use of a risk process.

The main principles in managing risk are:

- Identify and understand the main risks.
- Understand the impact and whether the risk is controllable or uncontrollable.
- Analyse the risks using an analytical technique, such as Monte Carlo simulation. Keep the analysis as simple as possible.
- Determine the attitude, capacity and appetite for the risks. Despite the theory, few organisations are risk loving unless the returns justify the risk.
- Allocation/apportionment of the risk through insurances, absorption and contract clauses passing the risk to another party, ensuring that they have the capacity to handle it.
- Produce a risk register explaining whether the risk is high likelihood high impact or low risk likelihood low impact.
- Embed the risk process in all decision-making.

The only certainty in life is that we will face uncertainty. Project teams are often blinded by 'illusions of certainty' and over-optimism; they also overestimate their abilities and underestimate what can go wrong.

Typical risks in construction projects are exceptionally inclement weather causing major disruption to programme, productivity rates on-site, getting paid, disruption

New Code of Estimating Practice, First Edition. The Chartered Institute of Building.

caused by labour disputes and suppliers and specialty contractors ceasing trading during the production phase. Bidding at low margins is unrealistic, with optimistic assumptions around cost, programme and procurement savings and inadequate provision for risk. Poor design information at the bid stage, caused by pressure on professional fees and client's seeking faster delivery times, creates large risk for the bidders. The client may have saved on design fees but increased the cost through higher bid premiums being added because of the unknowns.

The New Code of Estimating Practice (CoEP) is not intended to give detailed guidance on how to undertake risk management. Reference should be made to other risk management publications.

Risk management decisions stem from subjective judgements involving people and therefore moral and ethical issues as well as the technical, financial and physical issues. The reality is that failure is the biggest risk, either failure of an investment, process, product or operation. It is better to be concerned about the possibility of failure, that is knowing what might go wrong and taking adequate precautions, than knowing the mathematical probability that something may or may not occur.

The risk assessment tools that are available are based on scientific methods and structured procedures. They often, take little or no account of qualitative aspects such as risk perception or the human element. The tools deal in the science of probabilities, not the likelihood of human reaction in a particular set of circumstances.

Construction is high risk from the outset. If the estimate and bid are wrong, no amount of process improvement can change a poorly priced project into a financially successful project. There is enormous risk at the bid stage caused by larger and more complex projects, design ambiguities and the multi-layers of specialty contracting.

For example, the risk matrix in Fig. 12.1 only measures the likelihood and severity of each risk identified by the bid team for, in this case, trench excavation.

Construction projects have an abundance of risk, contractors cope with it and owners pay for it, and when things go badly wrong, the insurers and bond issuers also pay. Traditionally:

- The client/owner is responsible for the investment/finance risk and operating and maintenance risk.
- The design team is responsible for the design risk and sometimes the performance risk; they identify the risk and endeavour to control it, recognising that some risks are uncontrollable, such as the weather.

Hazards		Likelihood			Severity					Risk score
Ref	Key hazards associated with, say Trench excavation	Frequent	Occasional	Unlikely	Catastrophic	Critical	Serious	Marginal	Negligible	Likelihood x severity
1	Contaminated ground		✓			✓				
2	Underground services	✓				✓				
3	Site access	✓					✓			
4	Impact on adjacent properties		✓				✓			
5	High water table		✓				✓			
6	Trench collapse		✓		✓					

Very high risk | High risk | Low risk

Figure 12.1 Risk matrix.

- The contractor and specialty contractors are responsible for all aspects of the construction risk, including financial, health and safety, performance and time. Safety and health has become paramount in importance.

The management of risk is a key thread that has been taken through the CoEP to emphasise the greater responsibilities placed on contractors.

Even before the bid team is put together, there are risks that need to be identified and managed, such as:

- Competitive risks
- Financial risks (e.g. cost of bidding and getting paid)
- Legal risks (e.g. covenants and legal agreements).

There is a risk of 'not risking' and the company becomes subject to risk paralysis.

Once the tender documents have been received, the bidding and project risks need to be identified. This can be done by the bid team who would be allocated the responsibilities for managing the risks in their area and looking for opportunities to reduce or transfer them. A risk mitigation plan not only highlights the risks but also records their likelihood of occurrence. Risks can arise from the project, the design team and the contractor's own estimating process. Figure 12.2 shows the four steps to risk management.

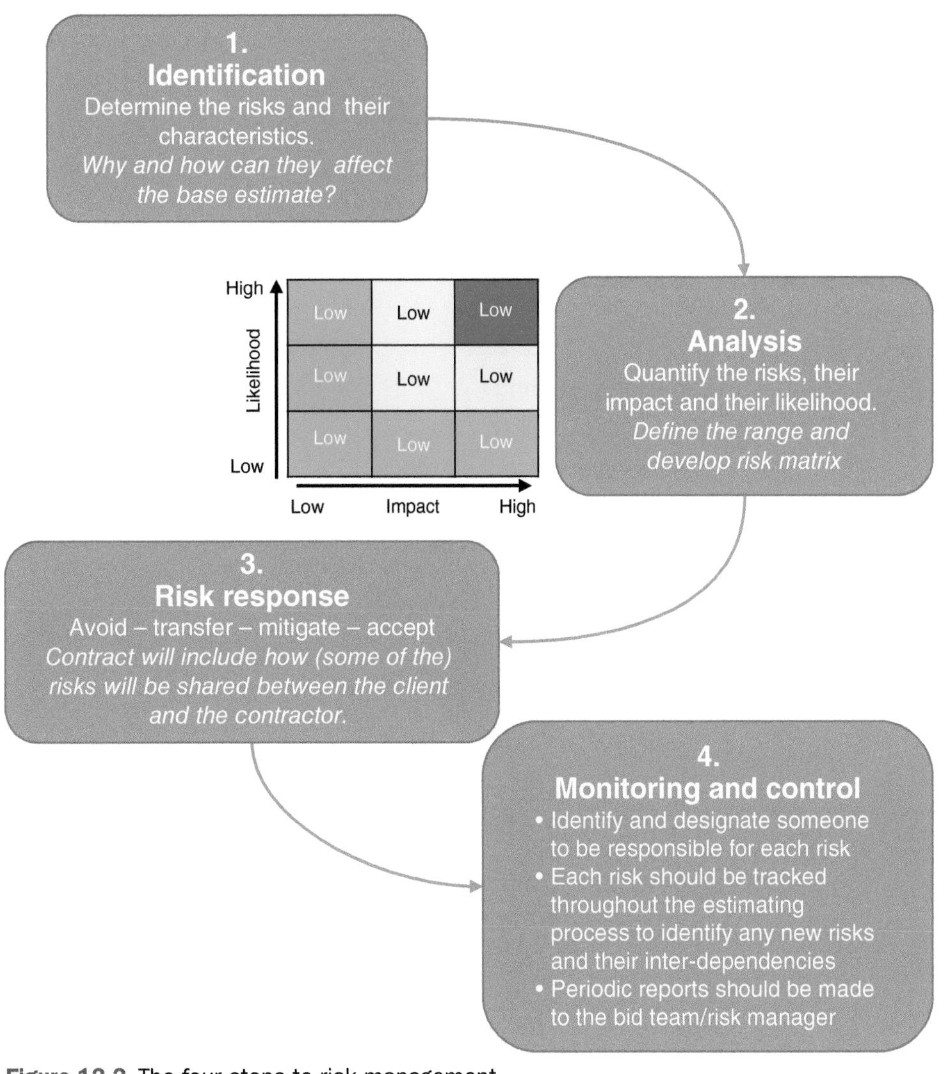

Figure 12.2 The four steps to risk management.

A tender submission needs to be developed in such a way that the risk mitigation and opportunities may be secured. For example, if a contractor assumes that a new gas tank can be sited at ground level on the site, they will clarify this as a condition of the offer and so pass on any risk to the client of any future requirement to bury the tank.

The design process has become more complex; CAD systems and building information modelling (BIM) offer great opportunities, many claiming it reduces uncertainties. However, the challenge is in integrating the systems and sub-systems. The bidding process has changed very little over many decades. Projects are bid using systems that relate to paper-based estimating, rather than complex digitised systems. Much of the risk inherent in the construction process stems from poor estimating and poor documentation at the bid stage, with the result that projects lose money for contractors.

The brief site visit at the beginning of the bid process to assess site location and constraints, which affect (and increases the risks in) the decision to bid. The uncertainty is increased at the early decision-making stages if the design brief is incomplete. Estimators tend to deal with the uncertainty by adding a contingency sum, usually a percentage of the estimated price.

This will often depend upon the estimator's risk attitude – see ***Risk allowance*** section. While there is widespread use of risk registers and brainstorming sessions, quantifying risk is still dominated by estimator bias.

12.2 Risk analysis

Figure 12.3 shows the process of gathering information to feed the risk register and the two different types of analyses – quantitative and qualitative.

12.3 Cost estimating accuracy

The level of risk in the project will impact the accuracy of the cost estimate. Project definition increases over the life of the project and, with that, there is a corresponding increase in accuracy. This is often called the 'estimating funnel' – see Fig. 12.4.

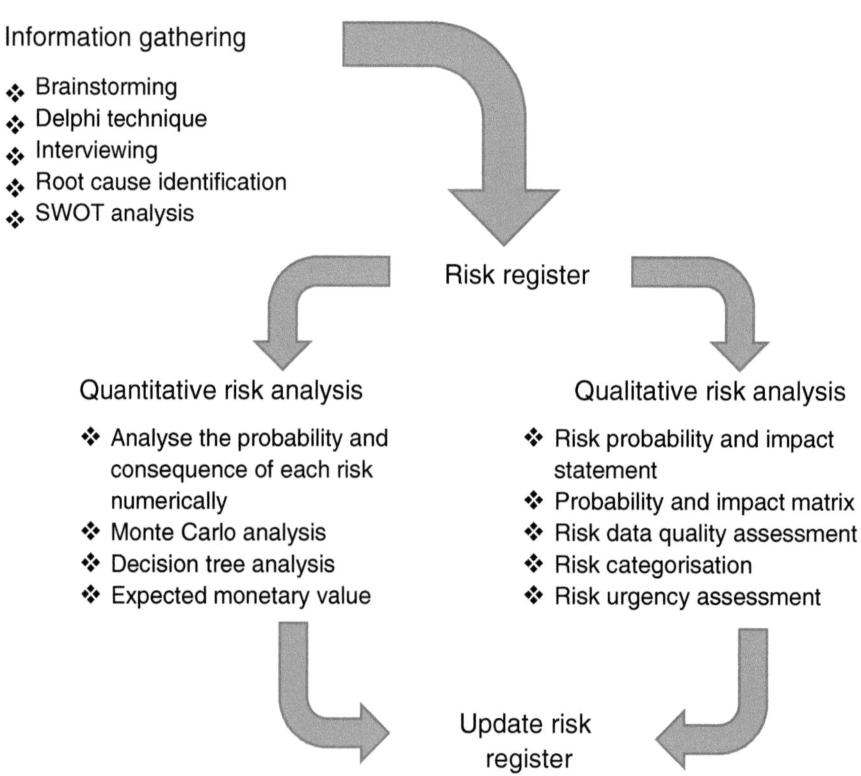

Figure 12.3 The process of 'feeding' the risk register and updating it.

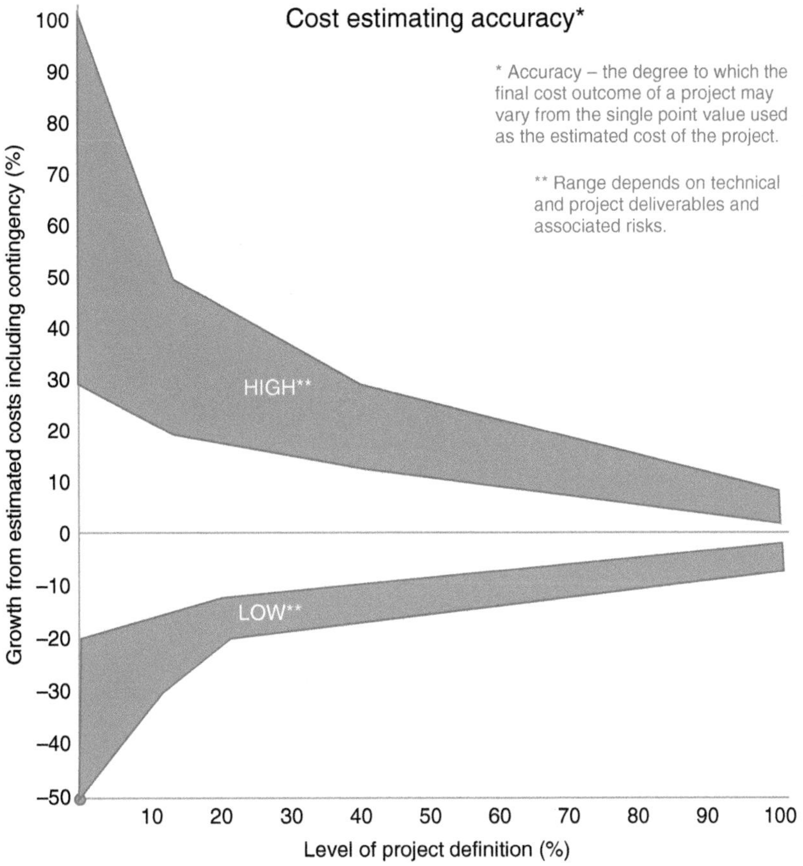

Figure 12.4 The estimating funnel.

13 Logistics

The underlying logistics plan for a project site needs to be understood in detail at an early stage. Otherwise, contracts, preliminaries, temporary works, cost estimating, risk assessment and health and safety will all be, at best, incoherent and incomplete. Construction logistics has become increasingly important and more specialised for medium and large construction projects. It is concerned with three main areas: project, site and supply. Underlying these are the gains from a good logistics plan, that is financial, environmental and customer satisfaction. The six principal areas are shown as hexagons in the logistics honeycomb.

The word 'logistics' is derived from the Greek adjective logistikos meaning 'skilled in calculating'.

In reality, logistics envelops much more, which is shown by the second honeycomb with its interlinking hexagons. This shows that integration/connectivity of the processes involved is key.

Logistics are important for organising people, information, materials and equipment and, in construction, this can mean releasing skilled operatives from being involved in unskilled work such as unloading deliveries and moving them around on-site. It would also avoid lost time due to waiting for deliveries and collecting materials, tools and equipment.

A definition of logistics includes the procurement transportation of personnel, a reference back to the original military context of logistics – 'the art of movement and quartering of troops'.

The key logistics process areas – project, site and supply (see Fig. 13.1). Each of these is underpinned (and interconnected) with three main pressures:

- Financial/resources
- Environmental (in the broadest sense, i.e. internal and external pressures)
- Client satisfaction.

In the main logistics honeycomb in Fig. 13.2, the hexagons have been grouped within the honeycomb by coloured lines to show their relationship with these processes – project (yellow), site (red) and supply (green), although the divisions cannot be clear cut as a number of the site issues impact supply and vice versa.

Poorly thought-through design can impact logistics and lead to unplanned temporary works and waste of time and resources. Therefore, early contractor involvement could have significant benefits. This highlights the link between logistics and value engineering, the latter being the identification and elimination of unwanted costs whilst improving function and quality. Value engineering should involve the whole project team in seeking cost-effective methods, solutions that add value to the

New Code of Estimating Practice, First Edition. The Chartered Institute of Building.

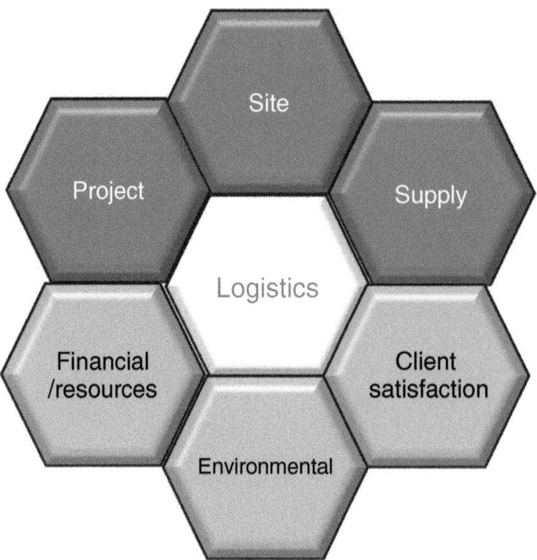

Figure 13.1 The key logistics process areas.

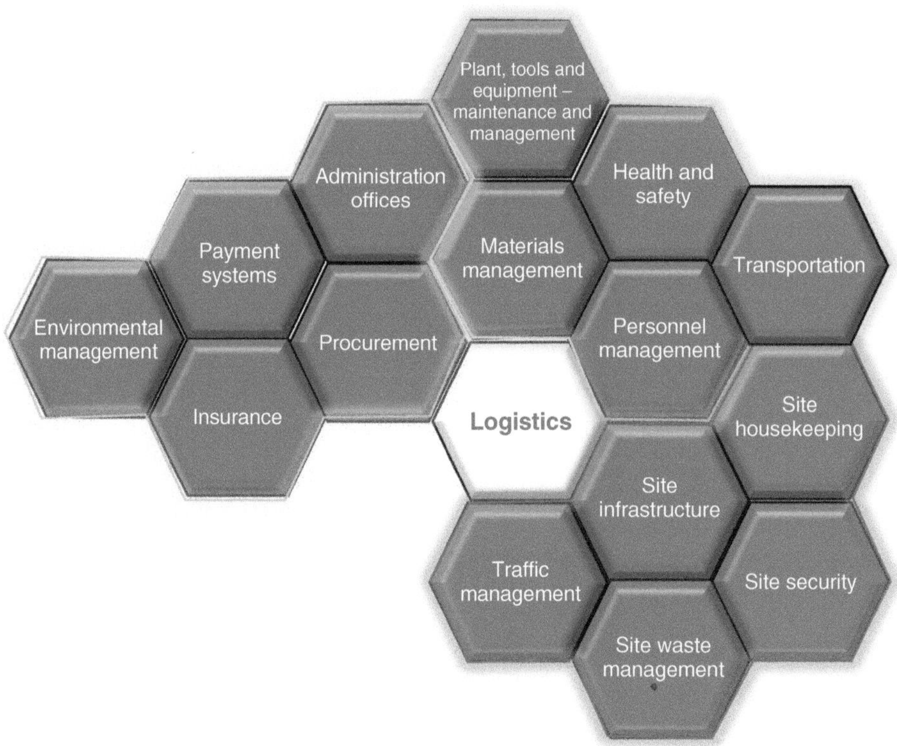

Figure 13.2 The logistics honeycomb.

project in not only time, cost or quality but also sustainability - see the section on *Value management/value engineering/value analysis*.

13.1 Materials logistics plan

The use of a materials logistics plan (MLP) can drastically reduce material wastage – a 35% reduction according to the waste and resources action programme (WRAP). Materials logistics planning is 'a practice designed to assist construction projects in proceeding smoothly whilst achieving programme certainty and cost predictability on complex building projects. [It] relates to the proactive management of the types

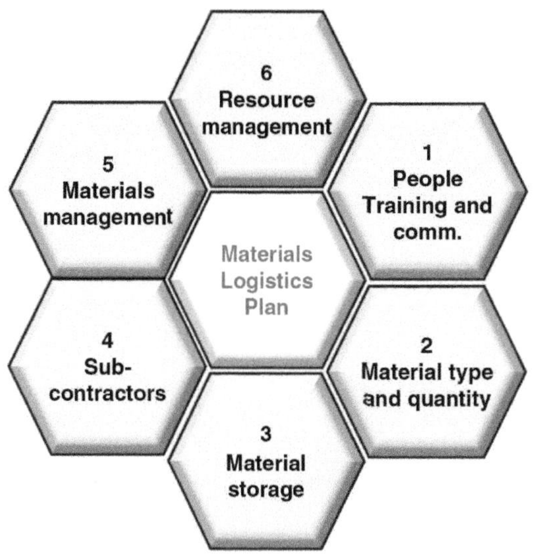

Key Facts/Main Benefits of logistics:

- *15% reduction in waste materials*
- *95% improvement in delivering performance (right materials, right place, right time)*
- *68% reduction of vehicles travelling to site*
- *25% reduction in accidents/injuries*
- *47% increase in site productivity*
- *75% reduction in CO_2 emissions*

Source: Wilson James (2015)

and quantities of materials to be used, including supply routes, handling, storage, security, use and reuse, recycling and disposal of excess materials' (WRAP, 2007).

1. *People – training and communication*: It is important to identify the person/people responsible for maintaining and enacting the MLP for each of the production stages/processes. Appropriate training will help to maintain the effectiveness of managing the plan. Communicating the plan to the project team and incorporating any feedback is essential. More importantly, this must include communication with sub-contractors and suppliers.

2. *Material type and quantity*: The quantity and type of material required for each process is outlined in the work plan (method statement). This information needs to be in the MLP along with:

- Delivery method
- Delivery timings
- Design waste % (e.g. offcuts)
- Process wastage % (e.g. overordering because of minimum load size)
- Justification of design and process wastage
- Supply route.

The requirement for wastage figures means that the MLP would be more efficient if linked to the site waste management plan – see **Environmental Management** honeycomb.

3. *Material storage*: The delivery of materials may be restricted by a planning condition, so the MLP co-ordinator needs to be aware of this. A description of where and how the materials should be stored, including any safety requirements, should be included in the MLP.

4. *Sub-contractors*: The fragmentation of the construction process with large numbers of specialty and sub-sub-contractors working on-site has a huge impact on logistics. On one hand, the movement and working methods of the sub-contractors can affect the site productivity; on the other hand, resource availability (capacity constraints) and poor site conditions can impact the sub-contractors' productivity. The MLP co-ordinator should be aware of the relevant contract conditions.

5. *Materials management*: Materials management is a process of planning, executing and controlling the right source of materials with the exact quality, at the right time and place, suitable for minimum cost construction process. It is a major part of the logistics plan, especially as materials account for a high percentage of the project cost. The MLP co-ordinator needs to ensure that sufficient space and resources have been allocated for the receipt and storage of materials. The planning conditions (if any) for supply routes (see *Transportation*) need to be adhered to, and guidelines for the safe, secure and appropriate storage of materials need to be developed.

6. *Resource management*: Resources include, people, materials, plant and equipment. The tracking, maintenance, storage and delivery of plant and equipment as part of a MLP can have a significant effect on productivity. Consideration needs to be given to the plant needed for unloading and moving deliveries to where they are needed or to site storage facilities. The plan should link to the method statement/work plan and the health and safety plan.

13.2 Materials management

Materials wastage has a significant impact on the project costs; 10–15% of ordered materials are either unused or end up as waste (WRAP). Waste is difficult to measure, but it has a major impact on the profitability of activities.

13.3 Personnel management and health and safety

Today's personnel management, in terms of logistics, reflects these military origins in that the aim is to get the right people in the right place at the right time and with the right tools and environment. A focus on worker welfare links personnel management with health and safety. Maintaining productivity can be achieved by providing people with a safe work environment and avoiding musculoskeletal disorders where possible.

13.4 Plant, tools and equipment – maintenance and management

Details are needed of what plant, tools and equipment are needed, from whom they are to be sourced, when they are needed and when they should be dismantled and moved away. Maintenance is important if the plant, tools and equipment are to remain efficient and safe.

13.5 Transportation

Transportation is a key part of logistics and is particularly important as the transportation costs of materials represent a significant percentage of project cost, and transportation in general has a huge environmental impact.

UK Government planning guidance, Planning Policy Guidance 13 (PPG 13): Transport, has the following objectives:

- To integrate planning and transport at the national, regional, strategic and local level.
- To promote more sustainable transport choices both for carrying people and for moving freight.

The benefits of using a construction logistics plan

Local authorities and residents

- ✓ Less congestion on local roads
- ✓ Reduced emissions to limit the impact of freight transport on the environment and contribute towards CO2 reduction targets
- ✓ Fewer goods vehicle journeys lowering the risk of collisions
- ✓ Opportunity to reduce parking enforcement activity costs – more deliveries should use legal loading facilities so less traffic and parking infringements will occur
- ✓ Improved quality-of-life for local residents through reduced noise and intrusion and lower risk of accidents

For building developers and contractors

- ✓ Reduced delivery costs and improved security
- ✓ More reliable deliveries resulting in less disruption to normal business practices
- ✓ Time-savings by identifying unnecessary deliveries
- ✓ Less noise and intrusion
- ✓ Opportunity to feed into a CSR programme and ensure your operation complies with health and safety legislation

For freight operators

- ✓ Legal loading areas will mean less risk of receiving penalty charge notices
- ✓ Fuel savings through reduced, re-timed or consolidated deliveries
- ✓ More certainty over delivery times to help increase the productivity of your fleet
- ✓ Less journeys will reduce the risk of collisions involving your vehicles

Source: Transport for London; Mayor of London

Figure 13.3 Construction logistics plan.

- To promote accessibility to jobs, shopping, leisure facilities and services by public transport, walking and cycling.
- To reduce the need to travel, especially by car.

In order to comply with PPG13, The Traffic Management Act (2004) (TSO 2012) and the London Plan, Transport for London (TfL) have developed construction logistics plans (CLPs) and associated guidance to developers and local authority planning officials – see Fig. 13.3.

A logistics plan focused on transportation needs a certain amount of information such as the type (size, weight and specification) of the vehicles visiting the site as well as their likely routes. The safety of pedestrians and other road users in the proximity of the site needs to be taken into account. This information is needed from not just specialty contractors appointed by the main contractor but any other party that may bring goods into the site such as sub-sub-contractors. Just-in-time (JIT) deliveries will benefit the logistics plan greatly, both in terms of traffic and storage on-site. The plan should include:

- Development description and location
- The construction phase details
- Planning requirements/restrictions
- Delivery booking and scheduling
- Supply chain management
- Off-site fabrication and consolidation.

Just-in-time (JIT)

JIT delivery was developed in the car-manufacturing industry where exact materials requirements could be forecasted to the minute. This has been developed within the retail industry through Point of Sale stocktaking systems and massive data analysis via loyalty cards. In short, JIT works against an exact forecast but the construction industry cannot do this. In reality, on a good site, materials requirements can be forecasted to within a week. It is not possible to have something arrive JIT if you don't know weeks in advance when that time will be.

This is the benefit of a Logistics Centre. It allows materials to be close but not on-site. Suppliers deliver in quantities and timings to suit their needs, which can be forecasted weeks in advance; then, the logistics sub-contractor delivers materials over the small remaining distance on a JIT basis because they are in constant contact with the site. A logistics centre removes the fundamental inefficiencies inherent in the incoherent nature of the manufacturing and construction industries.

Logistic/ distribution centres

A distribution centre for materials, plant and equipment can integrate the materials management, the plant tools and equipment and the transportation hexagons. This service, well established in the retail trade, is becoming more popular with contractors involved in complex projects and/or city centre locations where congestion and accessibility is a major problem.

Distribution centres can enable JIT deliveries of a wide range of supplies direct to the construction site, thus minimising the need for storage of materials. They also allow stock buffering, guarding against shortages or price escalations.

Their expertise in handling and delivering heavy plant, tower cranes and off-site prefabricated components can provide efficiencies and enhance project logistics.

The distribution centre can remove the packaging/pallets for later disposal/ recycling rather than taking up valuable site space. Secure, off-site storage can improve project costs and quality. However, it is not a warehouse for construction materials and components; the storage facility has a turnaround of approximately 10–15 days.

Instead, it can allow materials/components to be delivered in bulk for onward distribution, a huge benefit to a contractor with multiple sites in a relatively small area.

Materials can be consolidated to reduce the delivery of lots of part loads, which maximises fuel efficiency and reduces vehicle movements.

Logistics – site

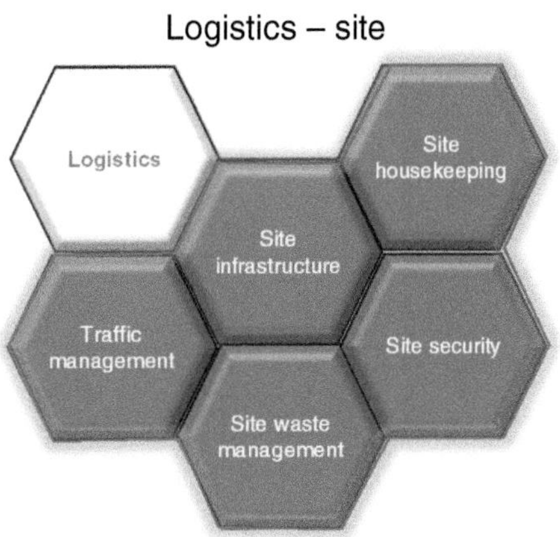

13.6 Traffic management

A traffic management plan (TMP) is a requirement of some planning applications. It is a good practice as part of the logistics plan, which may include:

- Development description, location and access
- Planning requirements/restrictions
- Routing of demolition, excavation and construction vehicles
- Neighbour consultations
- Scheduling
- Vehicle call-up procedure
- Impact on other road users
- Parking and highway licences
- Programme/key dates.

Site housekeeping

Site housekeeping issues such as quality, health and safety, waste and dirt generation can be improved by ensuring that materials are delivered to the site/work area/site storage precisely when they are needed and, at the end of a shift, unused materials and packaging are removed for recycling or reuse – see ***Waste management*** honeycomb under ***Preliminaries***.

Site infrastructure

This category includes drainage (temporary and permanent), water, electrics and compressed air supplies. It is closely linked to site establishment – see ***Site establishment*** honeycomb under ***Preliminaries***.

Site waste management

How surplus resources will be managed on- and off-site. See ***Waste management*** honeycomb under ***Preliminaries***.

Site security

This covers secure storage on-site and the appropriate security/safety measures for stored materials and plant and equipment – see ***Security*** honeycomb under ***Preliminaries***.

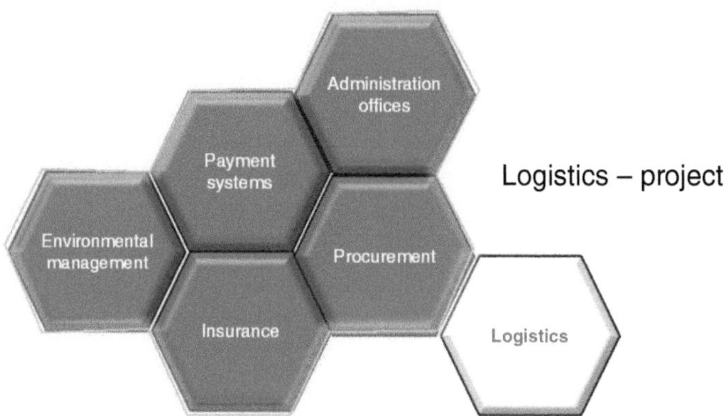

Procurement

A good logistics plan will ensure that the appropriate administration facilities/people are based on/or very close to the project site to avoid time wastage and travel costs.

Administration offices

Logistics can involve the management of procurement. Savings can be made by allocating procurement responsibilities across the company, reducing the opportunity costs associated by forming a new team for the delivery of each project (BIS, 2013).

Payment systems

Contractors'/client's payment performance can have an impact on specialty contractors' and suppliers' performance. Linking payment schedules to the logistics plan can support timely payments, thus improving the contractor's working capital and reducing risk.

Environmental management

A good logistics plan will include a link to the person/team responsible for environmental management of the project.

TfL's CLP requires a waste plan to ensure that waste collection is co-ordinated, fit for purpose and maximises any opportunities for reuse and recycling of materials on-site.

Insurance

Maintaining current insurance certificates and performance and payment bonds related to the project is important and can form part of the logistics plan and work of its co-ordinator.

14 Resource and production planning

Whilst the price can be built up by taking account of direct and indirect costs, overheads, profit and risk, there needs to be an allowance for time, for example if the client needs the work to be completed at a specific time of the year, or work needs to be undertaken in a very short period or completed in phases. Furthermore, finishing the work early or late will have a cost implication, depending on when practical completion is signed off by the architect.

Time also plays an important role in estimating productivity rates. Using historical data requires the estimator to take account of when the work was undertaken, the time pressures in place and external forces such as weather and climate.

The pre-tender plan must include a realistic project time frame. This requires a good understanding of the processes, people and products involved and careful ordering of the activities. These processes will rely on several sources of information, working with many different people in the organisation. The method statement is an essential part of the bid process as this describes the operations needed, the level of work involved and the appropriate time allowances. Alternative methods of work may also be shown for the estimator to consider.

Planning and scheduling are separate disciplines.

Project planning is largely an experience-based art, a group process requiring contribution from all affected parties for its success. On the other hand, scheduling is the science of using mathematical calculations and logic to predict when and where work is to be carried out in an efficient and time-effective sequence. It involves quantifying the programme. Planning must precede scheduling. They cannot be carried out in parallel, nor can scheduling precede planning (CIOB, 2011, p.1).

Planning requires decisions concerning:

- The overall strategy of how the work process is to be broken down for control into work packages or location
- How control is to be managed
- How design will be undertaken and by whom
- The methods to be used for construction
- The strategy for subcontracting and procurement
- The interfaces between the various participants

New Code of Estimating Practice, First Edition. The Chartered Institute of Building.

- The zones of operation and their interfaces
- Maximising efficient and the project strategy with respect to cost and time
- The management of risk and opportunity.

14.1 Planning techniques

Bar or Gantt charts are the simplest form of identifying the start and finish times of an activity. The times may be divided into days, weeks or months, and the activities can be at different levels of details, from work packages to the whole project. A number of assumptions are made in the creation of the bar chart. For example, that all activities will start and finish on the optimum dates and that productivity and resource availability are consistent. In reality, delays in activities and the delivery of materials or plant are all common occurrences on a construction project.

Flow charts can enable the introduction of logic to the activities identified in the bar chart. They help to manage processes and to analyse problems such as bottlenecks.

Critical path or network analysis takes account of interdependencies between the activities being scheduled. A network is drawn to show these dependencies showing the preceding and succeeding activities – see Fig. 14.1.

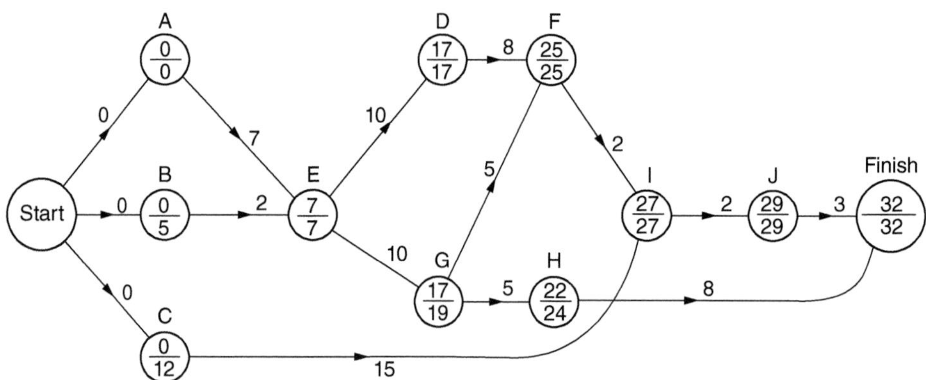

Figure 14.1 An example of a network diagram.

Where the top number in the circle is the earliest possible starting time, the number at the base of the circle is the earliest possible finishing time and the number on the arrowed line is the activity duration. The vertices with equal earliest and latest starting times define the critical paths: A E D F I J

14.2 Resource planning

Resources such as labour, plant, money and material need to be planned at the estimating stage, however approximate, to ensure that costs are covered. Time is also important, for instance in calculating labour or plant/equipment productivity. Duration is key and may not be calculable from data on productivity and resource alone. Reference to previous projects is a good source of information on productivity, but care should be taken that the comparators are realistic. The planned duration of an activity is a function of the quantity of work:

$$Duration = \frac{Quantity\ of\ work}{Productivity\ quotient \times Quantity\ of\ resource}$$

The resource planning honeycomb is made up of five hexagons, which are explained in the following sections.

Planning is 'the determination and communication of an intended course of action incorporating detailed methods showing time, place and the resources required'.
CIOB (2011)

Work breakdown structure (WBS)

Work breakdown structures are a good way of identifying the processes to be priced and can be used as the basis for the resource and time scheduling in the method statement. Effective scheduling will not only guide the estimating process but will also reduce stoppages/bottlenecks, late/untimely delivery of materials and help to meet the project deadline.

The work breakdown structure (WBS) approach is used widely in the US construction sector. It was originally developed by the US Department of Defense in 1968 and was mandatory on all their defence projects. The Australian Defence Department also insist on a WBS approach on their projects. It is an approach that is used in many industries including aerospace, oil, gas and petrochemical and pharmaceutical.

A WBS provides a common framework for all deliverables on a project and the specific tasks within that project. It encourages systematic planning, whilst reducing the risk of leaving out key elements in the cost-estimation process. This common framework helps in the communication of project details across stakeholders and thus improves integration in terms of time, resources and costs. In a construction context, the hierarchy for the WBS can be aligned to the BCIS elemental breakdown into different levels of analysis, for example:

- 0 Facilitating works
- 1 Substructure
- 2 Superstructure
- 3 Internal finishes and so on.

It can make up the composition of a Bill of Quantities (BoQ) as a BoQ breakdown structure (BQBS), which, in turn, initiates a cost breakdown structure (CBS).

These hierarchies represent a one-dimensional framework, but the estimator uses resources, time, sequence and method to build up a price. The framework needs to be three-dimensional showing the underlying risks and the connectivity with other work packages (particularly any clashes or pre-requisites) and the external environment. This necessitates a checklist integrated into the main framework that can be accessed by all those involved in the estimate. The framework also needs to record how the estimating structure was devised, the ground rules and assumptions made and allow for the corporate memory to be updated for future projects.

Organisational breakdown structure (OBS)

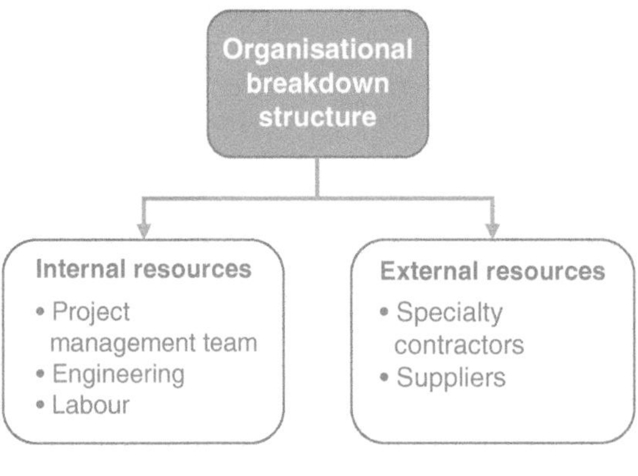

OBS is an hierarchical model describing the established organisational framework for project planning, resource management, time and expense tracking, cost allocation, revenue/profit reporting and work management.

Cost breakdown structure (CBS)

Costs are allocated to the lowest level of the WBS. The tasks at this level can often be subdivided into discrete activities to be completed by different departments; therefore, one task may have several costs elements.

Whilst productivity is closely linked to the crew size, the relationship is not linear; reducing resources, below or above an optimum level, can produce inefficiencies.

The productivity quotient can be determined in a number of ways:

- Published output rates
- Data from completed projects
- Advice from specialists
- Personal experience
- Benchmarking.

Importantly, a resourced schedule allows high-peak requirements or lack of continuity for trades or gangs to be taken into account.

Resource scheduling

Scheduling involves further decisions before being shared with the project parties, such as:

- When the work will be carried out.
- Its duration.
- The level of resources required.

The use of scheduling software is widespread from straightforward programs such as Microsoft Project to powerful pieces of software such as Primavera. The software allows the bar chart used in the scheduling to be interactive, allowing changes and updates that would not be available if a bar chart was produced.

Depending on the type and complexity of a project, there are a number of scheduling techniques. Whatever the technique, it is important to monitor events to reflect actual progress and the impact of any intervening actions. A simple bar chart of activities cannot meet these needs, it has no logic; a logic-linked activity network is needed.

The time model can be shown graphically through:

- Bar charts
- Line-of-balance diagram
- Gantt chart
- Arrow-diagram method
- Precedence diagram
- Linked bar chart.

More information about these different types of scheduling diagrams are in the CIOB's publication ***Guide to Good Practice in the Management of Time in Complex Projects (2011).***

Table 14.1 shows the items that should be considered when developing the schedule.

Specifications

The basis of the specifications comes from the tender documentation. The drawings should be reviewed to identify the main activities and elements. From this, the construction methods and schedules/durations can be assessed.

Table 14.1 A checklist of items to be considered when designing the schedule.

Time for completion	Licenses and permissions
Sectional and key completion dates	Provisional and prime cost sums
Unspecified milestones	Specifications
Access, egress and possessions	Bills of quantities
Information release dates	Environmental considerations
Submittals and approvals	Health and safety
Procurement strategy	Labour and plant resources
Materials' delivery and storage	Method of construction
Temporary works	Sequence of construction
Temporary traffic arrangements	Schedule requirements
Working hours and holidays	Updating requirements
Design responsibility	Notice requirements
Complexity of design	Reporting requirements
Adjoining owners	End-user requirements
Risk allocation	Testing and commissioning
Sub-contractors and suppliers	Furniture and fittings
Separate contractors	Phased occupation
Employer's contractors	Occupation and handover
Employer's goods and materials	Partial possessions
Nominated sub-contractors	Logistics
Utilities and statutory undertakings	Third-party issues

Source: CIOB (2011).

14.3 Time – its perception and impact on the estimating process

The perception of time in the estimating process is an important consideration. Time is a resource and, like any other resource in a construction project, it has limitations in terms of availability and cost.

At the bid stage, time is of the essence. The short timeframe between tender and bid submission puts pressure on the estimator. Time stress affects decision-making in a number of ways:

- A reduction in information search and processing
- A reduction in the range of alternatives and dimensions that are considered
- An increased importance of negative information
- Defensive reactions, such as neglect or denial of important information
- Bolstering of the chosen alternative
- A tendency to use a strategy of information filtration; that is, information that is perceived as most important is processed first, and then processing is continued until time is up
- Increased probability of using non-compensatory choice strategies instead of compensatory ones

- Forgetting important data
- Wrong judgement and evaluation.

Some decisions may be made very quickly and are described as heuristic, involving intuitive non-analytic decisions. This type of decision, in particular, relies on memory. Time is an important feature of memory in the recall of previous projects and their cost implications and past experiences.

In compiling an estimate, the estimator and others involved in the process will recall a limited number of items from memory. Items/instances that had higher impact on say cost, time or quality will be recalled more readily. Memory is both subjective and inconsistent, although the level of inconsistency will vary between people and across age ranges.

Research has shown that 'duration neglect' is a common feature of memory. For example, the estimator (or bid team) may remember the problem with a particular type of curtain walling but may fail to recall the period of time that had an impact on the project or how much extra labour time was needed. This is described as memory built up of a number of snapshots rather than the whole film (Kundera, 2000).

Recording 'lessons learned' and information from past projects, including what, how, why and when decisions were made, will reduce hindsight bias and reconstructed memory.

15 Computer-aided cost estimating

There are increasing demands on estimators, particularly with changing procurement methods, to provide more information for complex tender processes. Many tender documents are required to include a wider range of information (Hackett, 2010; CIOB, 2009) that demonstrates:

- Best value
- Design and buildability

and provides:

- Value engineering
- Cost planning and 'design-to-cost' exercises including whole-life cost assessment
- Risk management information
- Information on health and safety issues
- Supply chain management information
- Project duration.

These requirements mean greater sophistication in the estimating process and the management of large amounts of data. The use of computer-aided estimating is increasing as more software become available and online/electronic services are available from the supply chain. The use of computers can range from straightforward spreadsheets to integrated systems that span the estimating process. The laborious calculations are undertaken by the computer, reducing the contractor's time and mathematical errors. The reports produced by the software packages provide electronic records that can be shared, saving on printing and hard-copy storage costs. They are easily available for reference for future projects, saving time and money in the long run. Modifying/updating the stored information can produce cost-effective estimates.

The databases supporting the software typically include wage rates for different workers, productivity of components, plant and equipment. Automatic quantity take-off can save huge amounts of time and improve accuracy. Some of the packages allow estimates to be created from scheduling tools such as Primavera and MS Project; the calculation of productivity and 'what if' analysis and the development of estimates based on work breakdown structures providing a more formalised approach with greater possibilities for comparisons of historical data.

New Code of Estimating Practice, First Edition. The Chartered Institute of Building.

Builder's merchants offer online quote systems, including price negotiation. Many of them provide an estimating service.

Exporting information to other software packages such as scheduling and cost control can help to integrate and improve the process. Importing and exporting information is an important function of any software, but it can raise the challenge of interoperability. 3D models provide parametric objects to which additional information such as material properties and costs can be attached and form a contemporary cost-estimating technique (Zanen and Hartmann, 2010). The model can be linked to cost-estimating software, and there is an ability to update information in the model and all the processes linked to it.

Computer estimating systems must deal with 2D drawings and 3D/BIM models. The systems can generate automatic quantities from 3D models as well as producing quantities from 2D systems. Areas, lengths and all the basic measurements can be exported into an estimating system. Major design packages such as Revit®, Archicad®, Microstation®, Tekla® and Sketchup® have revolutionised the way design is undertaken.

Paperless estimating has become a reality, with take-off, pricing and managing information digitally. The automatic take-off and pricing leads to time savings.

16 BIM and the estimating process

16.1 Overview

BIM is the process and technology for producing, managing and sharing physical and functional data of a facility in a collaborative environment using digital representative models throughout project lifecycle processes.

Building information modelling (BIM) offers the potential to transform many of the processes in cost estimating. It has the capability to provide a shared digital resource from the early design stages to facilities management and improve the accuracy of project cost information. Whilst BIM is more commonly associated with visualisation, one of its strengths is in model-based estimating.

BIM is particularly useful at the bidding stage when design (and other) information is incomplete, and there are numerous changes as the bid is developed and more information is received from specialist and sub-contractors. Managing these changes electronically, which can be shared amongst the bid team, is an advantage. However, not all costs can be obtained directly from the 3-D model. BIM does not necessarily reduce the cost of the estimating process, but it can enable the basic processes to be undertaken electronically, leaving the estimator to carry out more complex and intuitive activities.

BIM provides a visual database of building components that could assist in:

- Quantification
- Collaboration
- Communication.

Quantification

BIM reduces the amount of work involved in take-off quantities, allowing dimensions and other data to be available by a simple click to generate quantities. Furthermore, if any changes are made to the design, the related information, such as number, materials and schedules, change automatically. This automatic change management and the ability to take off quantities electronically helps to reduce human errors. Accuracy of the estimate can be improved as quantities are linked to the cost database.

New Code of Estimating Practice, First Edition. The Chartered Institute of Building.

Collaboration

The digital platform of BIM allows a wide range of users to be involved and collaborate on the design and production process. This co-operation is particularly important for the estimator who needs to collect and collate information from many different people/sources to build up the estimate.

Communication

BIM's visualisation capability makes it easier to communicate with non-cognate stakeholders, such as participants in the client's organisation, using graphics instead of text.

16.2 The challenges of using BIM in the estimating process

To be effective in the cost-estimating process, BIM needs to satisfy a number of criteria. The antitheses of these are often put forward as some of the barriers to adopting BIM:

Interoperability: the BIM software needs to import and export processes with no loss of information and, where possible, be compatible with the estimator's existing file formats.

Visualisation: this can help in the production of conceptual estimates. It can improve navigation around the objects or building elements and provides a clear visual representation for those in the process unfamiliar with the estimating concept.

Efficient quantification process: this allows flexibility in choosing which information to take off. The tool should have the ability to perform partial or full quantifications.

Reliability: object properties or data need to be extracted from the BIM model with minimum loss of information or object properties.

Built-in categories/classification should be easily customised: it should be straightforward to customise the in-built standard measurement rules to correspond with existing measurement standards, such as NRM in the United Kingdom.

Change management: any changes need to be recorded accurately, and the model needs the ability for comparisons of previous versions to be made.

The generation and export of reports: reports need to be exported in a user-friendly format that coincides with their existing report formats.

Section Two
Processes – the practice

The Processes section has been divided into a series of honeycombs grouped together according to the estimating timeline. The process follows from pre-qualification to handover.

The figure overleaf shows the sequence of the estimating process for producing a bid for a tender. It flows from the pre-qualification process, evaluating the viability of preparing the bid, through putting the bid together by considering the resources required and the pre-production and planning process.

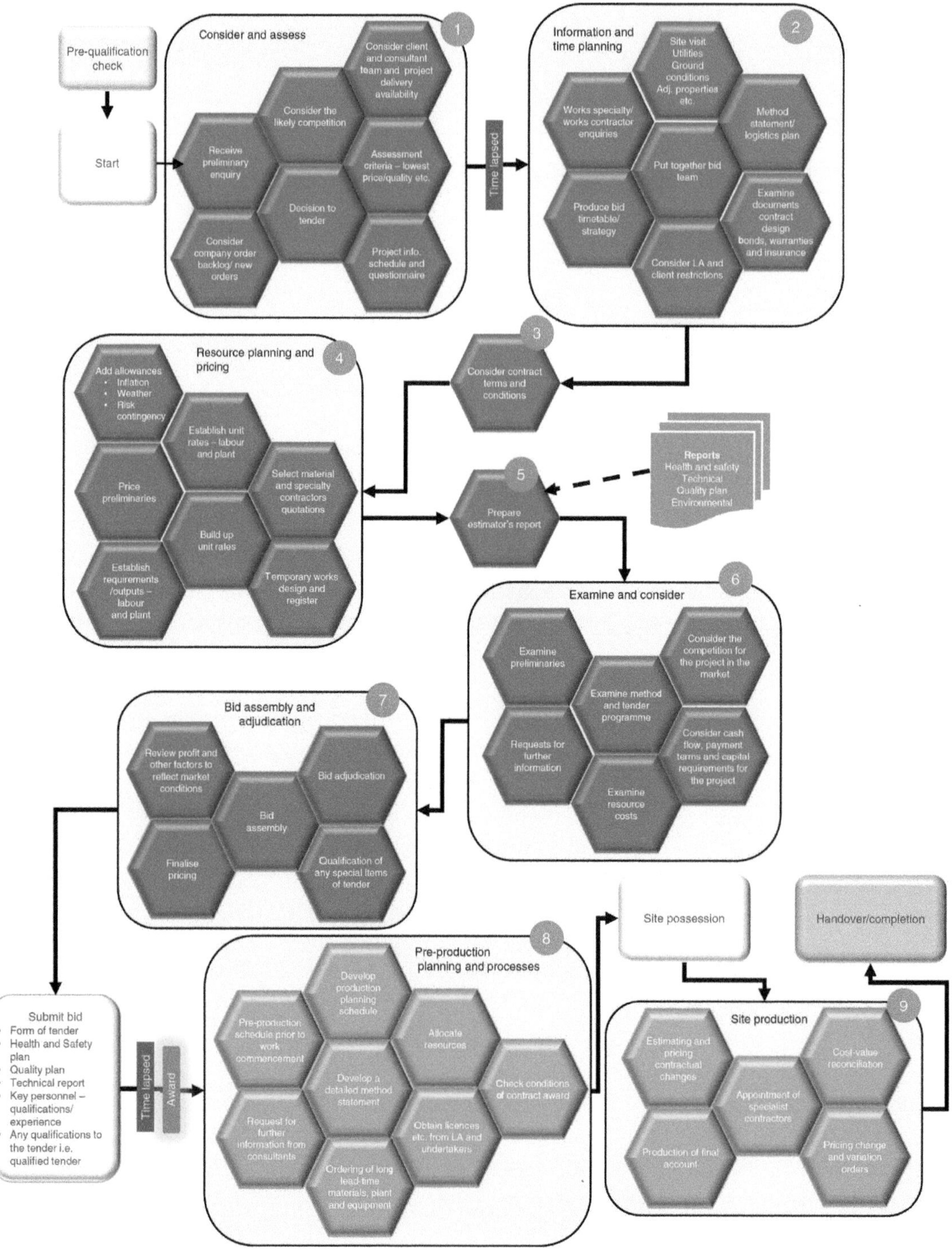
Pre-qualification check
Start
Consider and assess
1
Receive preliminary enquiry
Consider the likely competition
Consider client and consultant team and project delivery availability
Assessment criteria – lowest price/quality etc.
Decision to tender
Consider company order backlog/ new orders
Project info. schedule and questionnaire
Time lapsed
Information and time planning
2
Works specialty/ works contractor enquiries
Site visit Utilities Ground conditions Adj. properties etc.
Method statement/ logistics plan
Put together bid team
Produce bid timetable/ strategy
Examine documents contract design bonds, warranties and insurance
Consider LA and client restrictions
3
Consider contract terms and conditions
Resource planning and pricing
4
Add allowances
• Inflation
• Weather
• Risk contingency
Establish unit rates – labour and plant
Price preliminaries
Select material and specialty contractors quotations
Build up unit rates
Establish requirements /outputs – labour and plant
Temporary works design and register
5
Prepare estimator's report
Reports
Health and safety
Technical
Quality plan
Environmental
Examine and consider
6
Examine preliminaries
Consider the competition for the project in the market
Examine method and tender programme
Requests for further information
Consider cash flow, payment terms and capital requirements for the project
Examine resource costs
Bid assembly and adjudication
7
Review profit and other factors to reflect market conditions
Bid adjudication
Bid assembly
Finalise pricing
Qualification of any special items of tender
Submit bid
• Form of tender
• Health and Safety plan
• Quality plan
• Technical report
• Key personnel – qualifications/ experience
• Any qualifications to the tender i.e. qualified tender
Time lapsed
Award
Pre-production planning and processes
8
Develop production planning schedule
Pre-production schedule prior to work commencement
Allocate resources
Develop a detailed method statement
Check conditions of contract award
Request for further information from consultants
Obtain licences etc. from LA and undertakers
Ordering of long lead-time materials, plant and equipment
Site possession
Handover/completion
Site production
9
Estimating and pricing contractual changes
Cost-value reconciliation
Appointment of specialist contractors
Production of final account
Pricing change and variation orders

1 Consider and assess

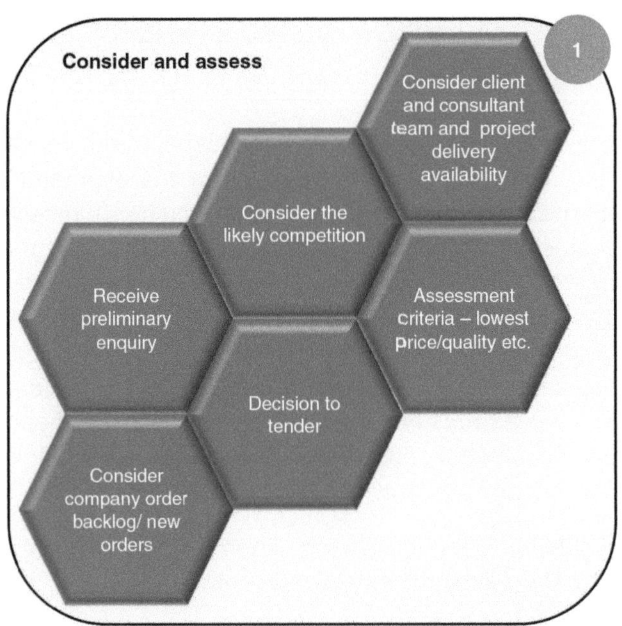

New Code of Estimating Practice, First Edition. The Chartered Institute of Building.

1.1 Receive preliminary enquiry

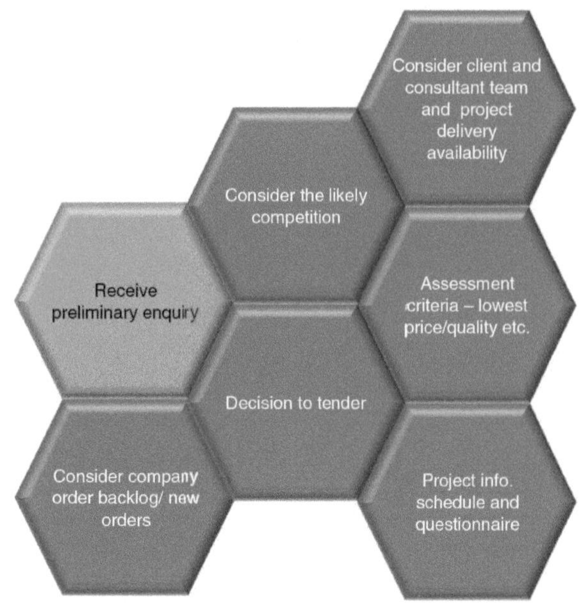

Time for bidding – The receipt of tender documents is the initial stage of the bid-to-tender process which usually takes 3–4 weeks for a traditional project, longer for a design-and-build project. There are many types of tenders (these are discussed further in the Principles section):

- Open tendering
- Selective tendering
- Negotiated tendering
- Serial tendering
- Framework tendering.

Bid documentation

An inspection of the tender documents must be made by the estimator responsible for the production of the bid and a checklist of the documents received. In large companies, the documents may be inspected by the chief estimator and other members of the contractor's organisation, including the planning engineer, project scheduler, quantity surveyor, buyer/procurement, contracts advisor and contract/project manager. Clear lines of communication ensure that all viewpoints of those examining the documents are considered. The estimator (or for larger projects, the bid manager) will be responsible for the co-ordination of these views. Inspection of the tender documents must seek to achieve, as a minimum, the following objectives:

- The documents and information received are those for the project under review, they are adequate for assessing costs and the risk
- Sufficient time is available for production and delivery of the tender
- The drawn information is sufficiently well developed to use for reliable pricing with no significant areas of uncertainty or unreasonable assumptions that must be made because of lack of information.

The tender documents on a traditional/conventional contract of design–bid–build usually comprise:

- Notice to tenderer
- Form of tender
- General conditions of the contract
- Specification of the works
- Drawings that form the basis for the tender
- Bills of quantities produced by the independent cost consultant appointed by the client.

The quality of the documentation must be assessed, thus avoiding lack of information and unreliable information which could lead to contractual claims at a later date. Enough information, such as clearly defined elements of work, will reduce the level of intuition/guesswork/assumptions/inaccurate allowances for risk and uncertainty.

1.2 Consider order backlog/ new orders

A contractor's order book and order backlog of work to be completed is a good indicator of business health. For publicly owned companies, the level of new orders will generate investor confidence and improve the share price. These measures are relevant when considering a decision to bid, they reflect the workload and have an impact on resources – financial, manpower, materials, plant and equipment availability.

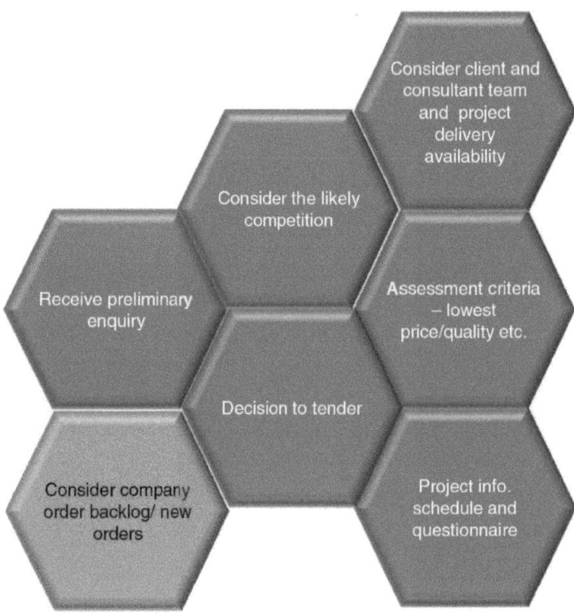

1.3 Decision to tender (bid or no bid)

Factors to be considered in the decision to bid:

- *The cost of preparing the bid ranked against the likelihood of success*
- *The time given in the tender to prepare the bid*
- *The workload of the estimating department*
- *Financial situation with cash flow and capital requirements*
- *Operational capacity with experience and competencies required*
- *Strategic direction of the business*
- *Any conflict of interest*

The decision to bid, the tender price and the level of competition are the major decisions in the tendering process. Many considerations have to be made in the decision to bid; each time the decision-making occurs, the circumstances, internal and external, may be different. By not submitting a tender for a project, the contractor may lose a good opportunity to make a good profit or to initiate/strengthen the relationship with a client or consultant.

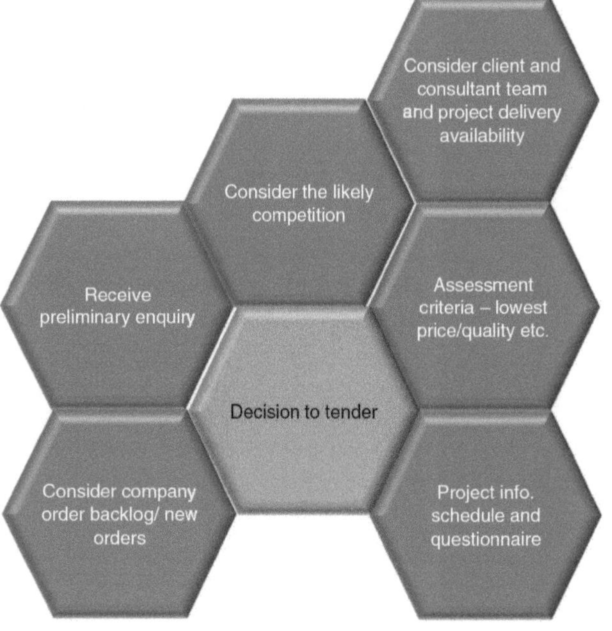

Three important influencers affect the decision to bid:

- Work in progress
- *Order backlog*: the amount of work in the pipeline which is underway; the usual measure is the number of months of work
- *New orders*: meaning projects where the contract has been signed, but work has not yet commenced on-site.

Deciding to bid on an inappropriate project could lead to financial and reputational loss. Submitting a bid costs money and involves risk. The cost of bidding must be absorbed in the company overheads. Success rates may vary; as a general rule, a success rate of one in six is considered an industry average. There is a need to balance risk and reward and to be aware of the opportunity costs* of any decision. Balancing workflow is an important activity for any company.

* A benefit, profit or value of something that must be given up to acquire or achieve something else.

Table 1.1 The factors considered by the contractor in the decision to tender.

Project type/ size	■ The location of the proposed works, including preliminary drawings and a site plan ■ Description of the project ■ Approximate price range of the project ■ The intended date for commencement/site possession ■ The stipulated period for completion, if stated in the documents ■ Details of any phasing and/or sectional completion ■ An outline of the form and type of construction ■ Site access issues ■ Special operational space requirements ■ Ground conditions, hydrology, soil type, borehole reports, risk of flooding ■ Sufficient dimensions and specification details to permit evaluation of the project ■ Details of work to be carried out by named sub-contractors, approximate value and details, if known ■ Indication of health and safety issues.
Client	■ Details of the client, or if a subsidiary company, details of the holding company, their credit worthiness and sources of funds ■ A description of the tender documents, their expected date of issue, the period available for tendering, the acceptance period for the tender and time when unsuccessful tenderers will be notified ■ Whether the project, either in its present or a different form, has been the subject of a previous invitation to tender ■ The latest date for receipt of acceptance of invitation to tender ■ The number of tenders to be invited (if available); this is not always adhered to in the private sector
Design team consultants	Full particulars of the consultants on the project, including their duties and responsibilities, and any experience of dealing with the consultants
Contract type	■ Form and conditions of contract ■ Modifications to standard clauses by the client ■ Liquidated and ascertained damages amount ■ Special clauses inserted in the standard forms of contract ■ Interim/stage payments terms ■ Retention conditions and release of retention ■ Any bonding requirements ■ Any stage completion dates ■ Details of insurance requirements ■ Provisions for fluctuations in prices ■ Warranties required.

The decision to tender can be made at one of the two stages:

- When pre-qualification enquiries are initiated by a client or their consultants, the contractor will make a decision based on an outline of the available tender information. The intention to submit a tender must be reaffirmed when the invitation to tender and supporting documentation are received.
- When pre-qualification is not a pre-requisite, invitations to tender can arrive without prior notice. In such instances, only one opportunity exists to appraise the project and make the necessary decision to tender.

Table 1.1 shows the important decisions in the decision to bid considered by the contractor.

In an ideal situation, there are four possible courses of action:

- Reject the project tender invitation

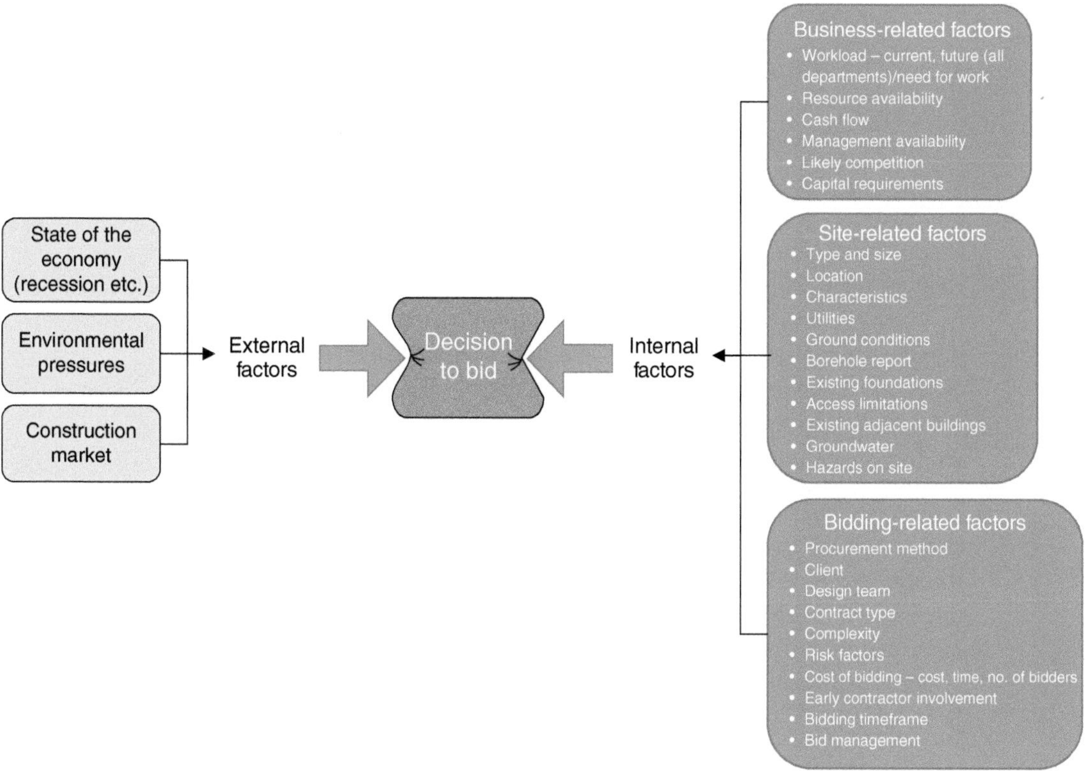

Figure 1.1 Overview of the factors involved in the decision to tender.

- Provisionally accept the project invitation to tender based upon particular requirements, such as time for bidding required
- Add the project to a reserve bid list
- Remove a project from the reserve bid list and replace it with the current project
- Unconditional acceptance of the invitation.

In reality, because of the cost and time incurred in tendering, the options may only be to accept or reject the invitation. If the contractor feels that rejection may prejudice future invitations to tender from a particular client, or consultant, then care must be taken to explain the circumstances of the refusal.

Many factors can affect the bid process – see Fig. 1.1:

- Internal (workload, availability of site team and financial)
- Project-related (experience, client, consultants, complexity, project programme and procurement approach)
- External (number and strength of bidders, market strength and competitors).

The estimating department may already have a full workload, hence the urgency in deciding whether or not to tender when documents are received. If the project cannot be accommodated because of workload in the estimating department, or the company's workload, the client must be advised at the earliest opportunity that a tender will not be submitted. This allows time for the selection of another contractor, if it is necessary to retain a full tender list.

Other factors

A. Procurement method

B. Workload (work in progress, order backlog, new orders)

C. Complexity

D. Risk factors

E. Bidding.

Procurement method

More information on procurement methods can be found in Chapter 7 of Principles section.

Procurement is essentially a series of considered risks – each has strengths and weaknesses. Different types of procurements (traditional/conventional, design and build, construction management, etc.) have strengths and weaknesses. Concession agreements, such as public–private partnerships, are more complicated with both costs and revenue streams that need to be considered at the bid stage. For some concession agreements, the contractor may have to purchase the concession by paying the project sponsor for undertaking the project.

Traditional/conventional contracting (design–bid–build), with the separation of design and construction using a lump-sum contract; the construction industry has used the traditional process for so long that it has become the most well understood. The understanding of the traditional approach is its greatest strength – the designer is responsible for design and the contractor for production, so responsibility for co-ordination of specialty works packages rests with the contractor.

Design and build is popular with clients; the design and production delivery risk primarily lies with the contractor. The process is relatively easy to understand – the project brief is specified (at least in part) and designed and built by the contractor, which, in theory, allows for better communication and integration between design and production. The design phase may be carried out by independent consultants hired initially by the client and then novated to the contractor. Estimating design-and-build projects takes longer and requires more work at the bid stage than the traditional approach. It must be made very clear at the tender submission what will be delivered for the price.

Prime contracting is an extension of the design-and-build system. The 'prime contractor' will be expected to have a well-established supply chain and will co-ordinate and project manage throughout the design and construction stages. The prime contractor is paid all actual costs plus profits incurred in respect of measured work and design staff – the main risk lies in respect of staff and the supply chain, including sub-contractors.

Private finance initiative/public–private partnership project (PFI/PPP) was launched in the mid-1990s by the UK Government, designed specifically for large-scale, high-value projects, such as road/rail infrastructure networks or hospitals, following the shift towards privatisation. For PFI, instead of paying a lump sum upfront for a new project, the Government agrees to pay a private firm an annual fee over a set number of years (the concession period) to take on the entire construction, financing, design, management and operation of the project.

Workload

See ***Consider company order backlog/new orders*** section.

Two separate decisions must be made concerning the workload:

- The chief estimator must make the decision based upon the workload of the estimating department and be satisfied that the estimator allocated to a project has the necessary expertise and knowledge needed for that particular task.
- Consideration must be given to the objectives and needs of the company, current and estimated future workload and the availability of resources to

undertake the project. Management must be satisfied that the project meets the company's objectives regarding type of work, workload and that the company is not exposed to undue risk by committing to excessive levels of work with a single client or one particular sector of industry.

Complexity factors

Complexity means complex processes and complex systems. A complex system consists of many inter-related elements. What differentiates construction industry procurement is the complexity of projects. Influences such as ground conditions, topography, design characteristics, logistics, weather, available technologies, finance, labour availability and services, just to name a few, all affect the ability of a project to be completed on time, on budget and to a high quality.

Risk factors

At the decision-to-tender stage there is a lot of uncertainty with many assumptions being made until time (and money) can be spent on clarifying the details. Risks need to be identified and assessed/analysed, with a view to how they will be avoided, reduced, transferred, shared or retained.

Risk allowances need to be included in the bid. Once identified, they are included with an allowance in the bid prices but may be adjusted at a later date at the tender adjudication stage by the management from a commercial/competitive perspective.

Risks can be separated into those to be borne by the employer and those to be borne by the contractor. In each category, they can then be classified as known risks, known unknowns (uncertainties) and unknown unknowns (*force majeure*). Risk allowances may be calculated on a percentage basis and include (RICS, 2012b):

- Design development risks
- Construction risks
- Employer change risks
- Employer other risks, which may include such issues as payment credit risk of the employer.

More information about risk can be found in the ***Risk Management*** section – Chapter 12 in Principles

Bidding

The accuracy of the bid; the type, size and complexity of work and workload are all 'internal' considerations. Bidding takes place in a much broader environment; the market and the competition are important factors that impact the tender.

The winner's curse— winning a project and failing to make the profit margin

Construction bidding is usually treated as a common value auction, that is the value of the item is the same to all bidders. Assuming that bids decrease with decreasing prices, the low bidder faces an adverse selection problem and only wins when they have one of the lowest estimates of the cost of construction. However, in construction bidding, each bidder will have a different estimate of the true value at the time they bid. Unless this adverse selection problem is accounted for in bidding, the low bidder is likely to suffer from a 'winner's curse', winning the project, but making below normal or even negative profits on some projects.

- *Suicide bidding*: The practice of bidding unusually lower than competitors in order to obtain work. Companies do this simply to ensure that they have work for their skilled staff to undertake, even if it means only breaking even on a project or, in some cases, making a loss.

1.4 Project information, schedule and questionnaire

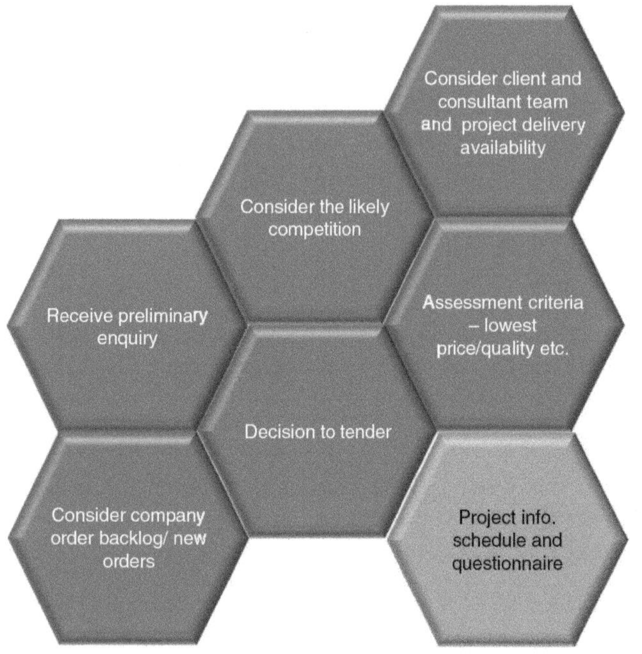

The following items need consideration following receipt of documents and the site visit:

- *Documentation*: Schedule of rates, bill of quantities, bill of approximate quantities, design and build and public–private partnership
- *Scale and size*: Scale of the project, amount and extent of specialty works packages, with their interdependence and which works packages have the potential to delay the works
- *Fixed or fluctuating price*: Fixed or fluctuating price contract
- *Design team and client team*: The client and the consultant team; experience of having worked with the team on previous projects. Reputation of the design team for collaborative working
- Duration stipulated by the client in the tender documents or the need to bid on project duration. Criticality of duration (speed of construction) in winning project
- *Extent of design completion*: Design development with details and the extent of the design information available at the tender stage. If there is still extensive design work to be undertaken, allowances should be made in the bid prices for potential disruption to the work schedule and the cost consequences
- Potential for changes to scope and design details that will lead to programme interruption and necessity for programme acceleration with specialty contractors
- Likely design changes to the project during the production process resulting from client changes and the need to develop design details.
- Complexity and the need to understand the complexity and intricacy of the design
- *Location and boundaries*: Site location, site boundary, site constraints, restrictions, accessibility of the site, rights of access, restrictive covenants, party walls, any special requirements and restrictions
- *Site survey information*: site survey, ecological survey, protected species and presence of any archaeological items that may delay the works
- Site investigation reports report on ground conditions, presence of difficult ground conditions and water table level in the area. Possibility of any flooding
- *Special employer's requirements*: Employer's requirements for sectional completion, special insurances, corporate social requirements for the local community, hoardings and advertising

- *Conditions of contract*: Contract terms and conditions that need careful attention, use of standard form of contract and modifications to the standard form
- The construction programme which, dependent on the contract, may be a formal contract document
- *Cash resources*: Cash flow, capital and credit requirements to undertake the project
- Stage payment times, retention amounts and timing of release, retention amounts on stage payments and practical completion, retention period and financial standing of the client
- Sequence and method of construction for the most efficient production
- Special safety and health issues on the project, such as presence of asbestos, handling hazardous materials and contaminated land
- Impact of weather on the work schedule sequence and construction programme; impact of winter on groundwork
- Market conditions affecting the award of specialty contracts and material prices. Consideration that market changes will affect commodity, labour and material prices
- Inflation allowances over the duration of the project if project priced as a lump sum fixed price contract. Likelihood of prices increasing for key items
- Possibility of discovery of any antiquities that could disrupt the project schedule
- Understanding the uncertainties and risk
- Amount of risk transfer† by the employer to the contractor and any unreasonable risks
- Extent of defined and undefined provisional sums‡ in the bills of quantities.
- Extent of controllable and uncontrollable items
- Requirement for alternative items for inclusion in the bid. The contractor must evaluate and price each risk transfer in the tender. There can also be risk sharing between the employer and the contractor and risk retention by the employer.

1.5 Consider client and consultant team and project delivery availability

Establishing integrity and trust between the contractor and the client/consultant team is critical to the success of any project. Understanding the client and their expected outcomes for the project is an important factor for a successful bid. The client, consultant team and the contractor share the same success factors:

- Clear project goals
- Clear definition and understanding of the project scope
- Clear understanding and appropriate allocation of risks

† The contractor must evaluate and price each risk transfer in the tender. There can also be risk sharing between the employer and the contractor and risk retention by the employer.

‡ A provisional sum for defined work is a sum provided for work that is not completely designed, but the scope and quantities of the work are defined in the bills of quantities. An undefined provisional sum is where the work is unknown and cannot be completely defined. Undefined work creates difficulties in producing the tender construction programme because of the unknowns.

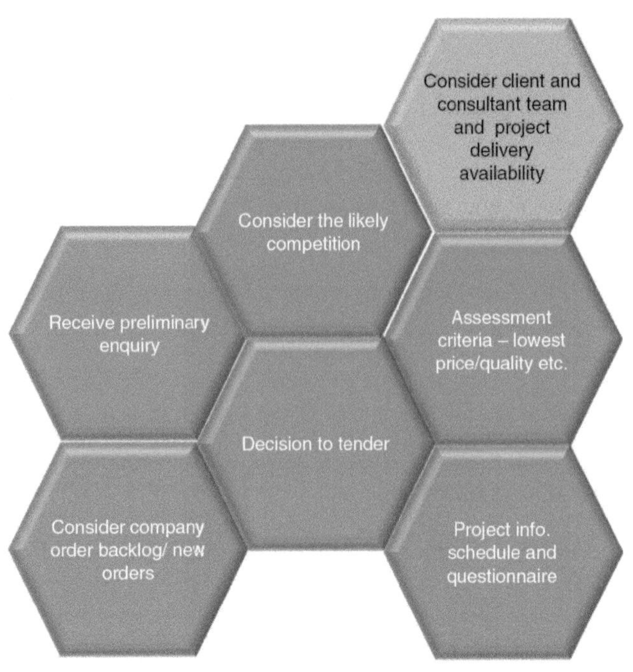

- Agreed risk/reward arrangement
- Appropriately skilled project staff
- Well-defined communications through all levels of the contracting parties with proper empowerment for decision-making.

Early contractor involvement

In two-stage tendering, the main contractor may be involved at an early stage, before all the tender information is available. It is a limited appointment and is often made with a specialist contractor. Early contractor involvement (ECI) may occur in a design-and-build tender, as the design requirements may not have been fully realised by the client. The tender at the first stage usually requires a method statement, preliminaries, the rates to be applied to the second tender, agreed fees and contract conditions.

ECI allows a closer relationship with the client as well as a better understanding of the project scope and goals.

Project delivery

Each project requires a certain set of competencies for its successful completion. The availability of those competencies/capabilities, either in-house or outsourced, is an important consideration in the bid process. Whilst labour, material, plant and equipment resource levels need to be considered carefully, the level and mix of skills to match the project requirements is also very important.

Consultant team

The consultant team is appointed by the client and works closely with the contractor on their behalf. Collaboration and communication is important between the consultant team and the contractor at the bid stage. Good knowledge of the team and previous experience of working with them before can be invaluable.

1.6 Assessment criteria – lowest price/quality and so on

LPTA The US Federal Government adopts a lowest price technically acceptable approach to some contractor selection.

Understanding the criteria that the client will be used for tender selection is important. It may be price, duration, quality, safety or a combination of any of these. Knowledge of the construction team and past experience can also be a factor.

Good procurement practice requires that all tenderers should be made aware of the bid assessment with any changes to these being communicated to ALL potential bidders to ensure parity of tendering.

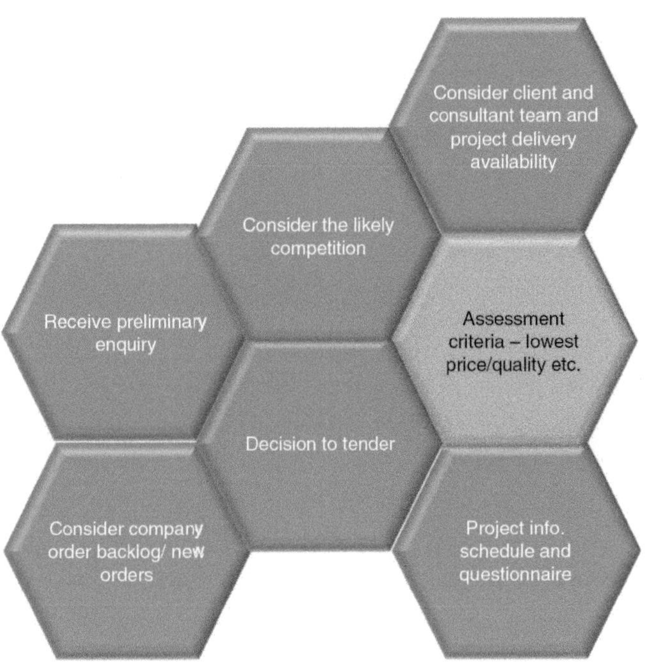

The most common factors are cost and quality, but increasingly complex projects and more innovative procurement methods have led to a larger set of criteria.

For example, the EU Public Contracts Directive (adopted in the UK's Public Contracts Regulations, 2015) states that the tender decision should be made on a 'most economically advantageous tender basis (MEAT)'. This can be on a price/cost basis only as well as other methods including best price/quality ratio, that is value for money.

Some of the criteria may include social and environmental requirements if they are related to the contract. The Directive places a duty on the authority procuring the work to investigate abnormally low tenders and disregard those in breach of international environmental or social law.

A MEAT offer uses criteria including:

■ Quality	■ Running and operating costs (if applicable)
■ Price	■ Cost effectiveness
■ Technical merit	■ After sales service
■ Aesthetic and functional	■ Technical assistance
■ Environmental characteristics	■ Delivery date, delivery time, and period of completion

1.7 The likely competition

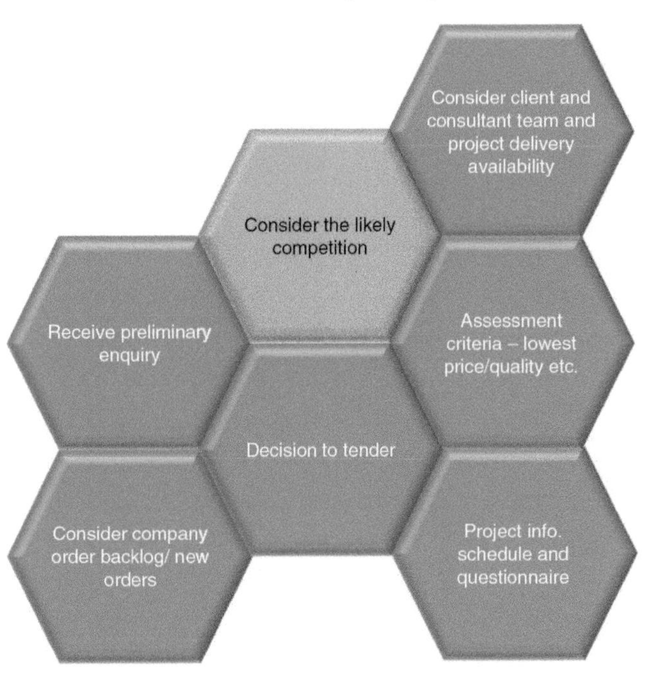

The numbers and quality of the competitors need to be taken into account. New competitors will need to be assessed for their bidding techniques, competencies and existing relationship with the client. With pre-qualification of bidders, the experience and capability of the bidders will already have been scrutinised. Consideration of the competition also needs to go alongside the benefits, or otherwise, of bidding for the project to maintain a link/relationship with the client.

Apart from the competition's capabilities/ experience, their workload should be taken into consideration. Any specialist knowledge (related to the project) needs to be taken into account, including experience in particular weather/climate conditions, knowledge or access to local labour/expertise. The procurement method will have an impact.

Where value, rather than price, is the decider, this can affect the reaction of the competitors involved in the tender. For example, PPP projects have high barriers to entry because of the cost of bidding, the capital required and the access to project finance. Similarly, design-and-build projects take longer to bid and involve high levels of risk for the contractor. Smaller firms find it difficult to compete for design-and-build projects because of the cost of tender submission.

2 Information and time planning

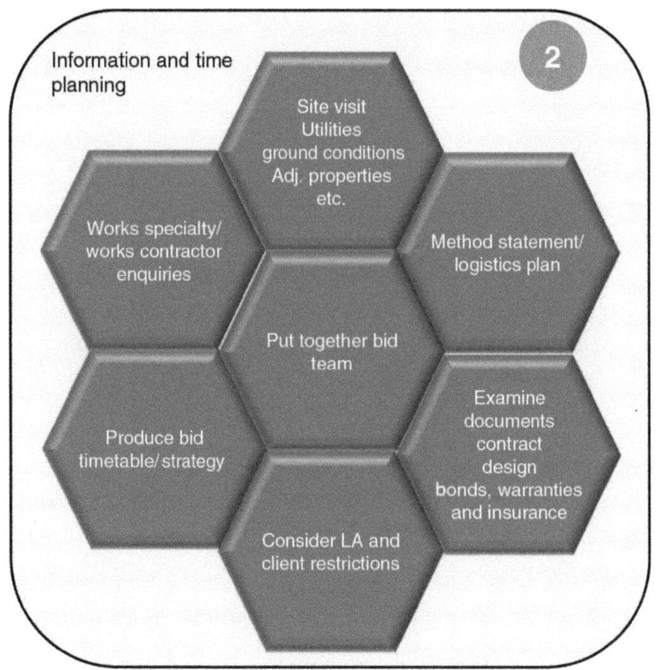

New Code of Estimating Practice, First Edition. The Chartered Institute of Building.

2.1 Works specialty/works contractor enquiries

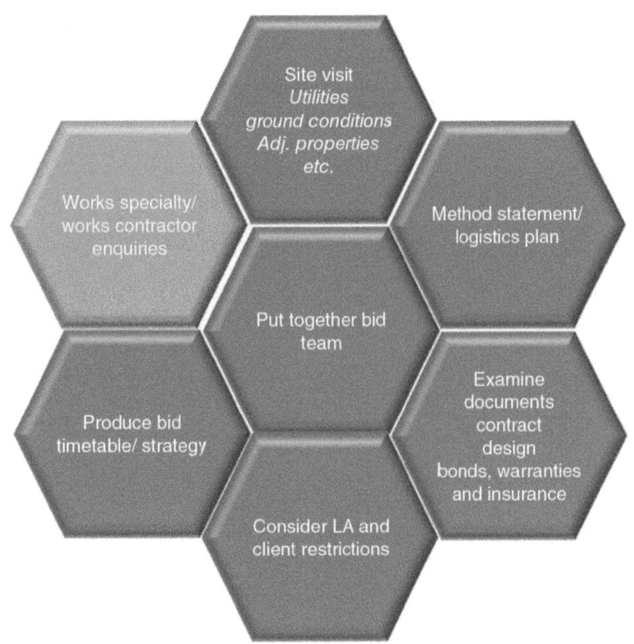

There are different types of specialty contractors, sometimes called sub-contractors or co-contractors:

- The conventional specialty contractor who provides a complete service
- Labour-only specialty contractors who are supplied with materials and plant by the general/prime contractor
- Labour and plant specialty contractors who receive their materials from the main contractor.

The key to success for any project is maintaining good relationships with the supply chain. They are also stakeholders in a project, and there must be respect, trust, prompt payment, co-operation and transparent and good communications.

It is now commonplace for contractors to 'single point' or even 'partner' with specialty contractors. Many contractors form a strategic alliance with their specialists. Such 'partnering' agreements typically give a contractor an undertaking by the specialist to provide an improved service and preferential prices in exchange for an undertaking by the contractor to give them a continuous flow of orders or enquiries. More sophisticated arrangements may incorporate an undertaking to make continuous improvements to performance and prices.

Innovation is an important part of bidding. Seeking an innovative solution on a bid can create competitive advantage.

Partnering agreements with specialists and with clients commonly incorporate key performance indicators (KPIs) that give a quantitative measure of performance in key areas such as:

- Quality of work or product in terms of finish and function and so on
- Achievement of programme objectives
- Keeping to cost targets
- Health and safety standards
- Training of staff and operatives
- Considerate contractor policies.

Consultants (architects, engineers, etc.) have traditionally been engaged and paid by the employer directly. The engagement of consultants is now often part of the contractor's procurement process and must be managed effectively.

Supply chain management

Contractors should maintain comprehensive records of suppliers and specialty contractors in the supply chain. These records include:

- Details of past performance on-site
- Previous performance in returning prices on time

Code of Considerate Practice under the Considerate Constructors scheme*
Considerate constructors seek to improve the image of construction by achieving best practice under the Code of Considerate Practice. The Code outlines five areas considered fundamental for registration:

1. *Care about appearance*
2. *Respect the community*
3. *Protect the environment*
4. *Secure everyone's safety*
5. *Value the workforce*

Construction sites, companies and suppliers agree to abide by the Code. After registration, a Monitor will review performance. The cost of registration will be included in general overheads.

- Extent of geographical operation
- Size and type of contract on which previously used
- Information concerning contacts
- Address, telephone and fax numbers
- E-mail address and website (if they have these)
- Notes on quality assurance (QA) registration.

When operating in a new geographical area, information concerning the local suppliers and specialty contractors is required. Performance should be verified from other external sources and any remaining information established from the supplier and sub-contractor concerned. A questionnaire may be used to establish the resources and abilities of specialty contractors/sub-contractors concerning:

- Area of operations
- Size and type of work
- Labour and supervision available
- Size and type of work previously carried out
- References from trade, consultant and banking sources
- Insurances carried by the specialty contractor/sub-contractor (if relevant)
- Confirmation of holding of relevant specialty contractor/sub-contractor's tax exemption certificate (if relevant)
- ISO9001 Quality Assurance Registration
- ISO14001 Environmental Standards and Registration
- Health and safety record.

This attempts to establish the supplier/specialty/trade contractor's compliance to undertake the work, with the resources, plant and equipment to meet the requirements of the main contractor's programme.

Pre-qualification is necessary when dealing with new suppliers and specialty/trade contractors. The pre-qualification questionnaire prepared by the contractor must ensure that the suppliers and specialty/trade contractors are compliant with the statutory, industry and company standards, with the necessary human and financial resources.

Selecting the right supply chain and developing good relationships is critical to winning a project.

Figure 2.1 shows the interdependence between companies, with the specialty contractors submitting quotations to more than one contractor bidding for a project.

Preparation of enquiry documents

A list of items for each enquiry package is made by electronically abstracting these from the tender documents. This list is used as a tick sheet while assembling the package, as a record of what enquiry documentation was sent and should be reproduced in the enquiry document (as a documentation log) so that the recipient can check that all the relevant components have been received.

* The Considerate Constructors Scheme is a non-profit independent organisation founded by the construction industry to improve its image.

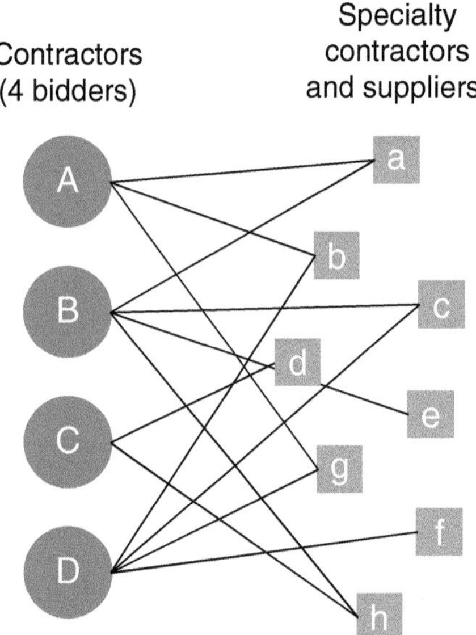

Figure 2.1 The interdependency between contractors and specialty contractors.

If bills of quantities are not provided with the tender documents, then the contractor must decide what quantities are to be prepared.

Recipients prefer to receive enquiries with bills of quantities; these require far less time and resources to price. There is a risk factor to consider in the absence of a bill of quantities, with the specialty contractor/supplier ensuring that the work content is measured correctly.

If no bills of quantities are provided with the enquiry, there should be at least a list of headings against which a breakdown of a lump sum price can be given. This will allow the estimator to compare and evaluate prices using the same list of headings and not have to base an assessment on the total price with no opportunity of interrogating it.

For larger projects, there may be an outline programme provided by the contractor; the specialty contractor would be required to work within the allotted time for their bid package, including any extended delivery periods which might impact the programme.

In design-and-build projects, specialty contractors have additional responsibilities, such as the development of the design for their works package and completion of working and shop drawings. They are also required to submit the risks identified and priced in their quotation – this ensures that the contractor does not duplicate risk allowances in the bid.

Under CDM (2015) regulations, requirements have been placed on efforts to establish competency, particularly where design responsibilities are involved (HSE, 2015b).

Scope of work

The documents sent out to a selected specialty contractor must clearly define the scope of the work, with as much detail as possible, with the sequence of working, the duration requirements and any restrictions on working times.

Enquiries for materials

Enquiries concerning materials suppliers should include:

- Title and location of the work
- Description of the materials, supported by specifications

- Quantities, so that bulk discounts can be quoted
- Date by which the quotation is needed
- Name of the estimator dealing with the tender
- The contract period with a guide to the dates for delivery
- Fixed price or fluctuating price
- Discount terms required
- Any limitations on access to the site
- Responsibility for unloading and storage of materials
- Value-added tax requirements.

2.2 The bid team

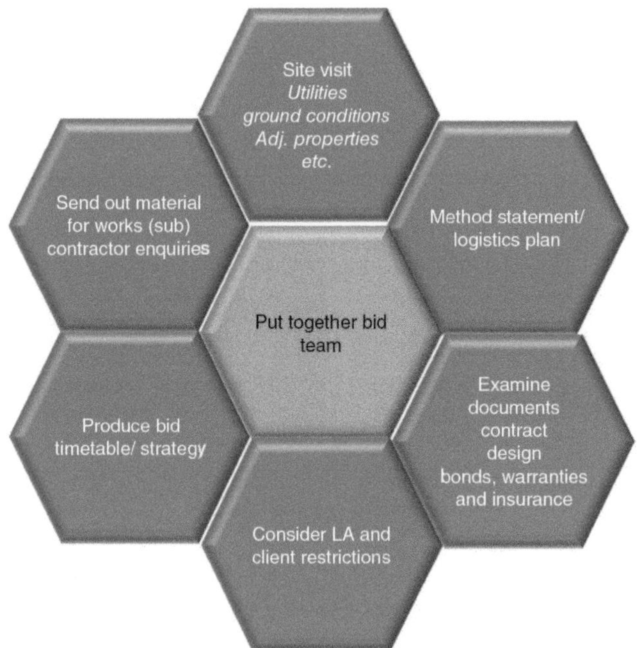

The team will embrace:

- Production methods including temporary works
- Planning the sequence of the work and scheduling resources
- Cost management
- Health and safety requirements
- Quality assurance procedures
- Logistics of supplies and deliveries
- Legal and contract requirements
- Purchasing and supply with long lead time delivery items
- Plant requirements
- Co-ordination of any client and local authority requirements.

The team must convince the client of the:

- Capability/competencies of the company, ensuring that the tender documents describe and reflect the competency of the whole production delivery team
- Credibility of the delivery team with the appropriate experience
- Reliability of the production team to meet deadlines and tight schedules
- Commitment to health and safety standards
- Commitment to quality control and quality assurance
- Policy on inclusivity and social responsibility
- Commitment to green policies on the environment
- Integrity/trust of the company
- Commitment to the project with key personnel.

The nature of a project-based industry such as construction means that teams will differ from bid to bid.

The experience of regular clients, their priorities, procurement approaches and evaluation methods are an important factor in knowing how the tender will be evaluated.

The bid team should develop risk management strategies for the bid process to avoid an unrealistic/over-optimistic bid, avoiding 'illusions of certainty'.

A bid manager is responsible for submitting a completed bid to an existing or prospective client, on time and within budget. All the client's questions must be answered as fully as possible, to give the organisation the best possible chance of success.

Communication across the team is crucial throughout the bid process, an important role of the bid manager. Some bids will be very straightforward, whereas some will be lacking in information and detail, requiring the tender to be qualified about the assumptions made.

2.3 Produce bid timetable/strategy

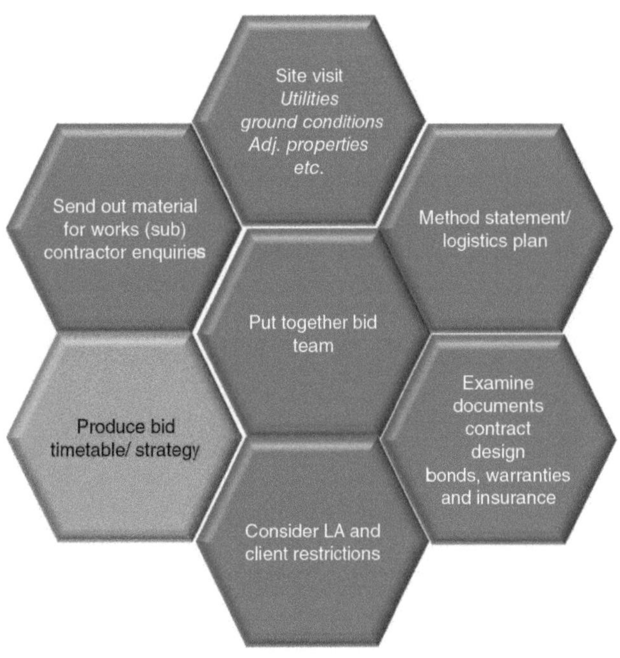

The bid timetable highlights the key dates in the production of the tender.

- Latest date for dispatch of enquiries for materials, plant and specialty contracted items
- Latest date for the receipt of quotations
- Key dates for bills of quantities production, drawings and specification for design-and-build projects
- Visit to the site and the local area
- Finalisation of the method statement
- Finalisation of the pre-tender construction programme
- Completion of pricing the measured rates
- Intermediate co-ordination meetings for the bid team
- Review meetings
- Submission of the tender.

Personnel associated with the tender must confirm that they are able to provide the necessary data in the format required, in accordance with the agreed timetable.

Tender preparation

The time available for the production of the tender should allow for the assimilation of project information, obtaining quotations from speciality/trade contractors and suppliers, and for completing the pricing.

The planning engineer and/or project scheduler should prepare a resourced network diagram/bar chart of the estimating activities, indicating the activities, durations, sequence and timing necessary for the completion of and submission of the tender. The pre-tender construction programme sets out the framework on which the detailed construction programme is based following any contract award.

Co-ordination with management and other departments within the contractor's organisation establishes key dates, decides on necessary actions and monitors progress during the production of the bid.

The bid preparation programme should show:

■ Who is involved	■ Tender submission date.
■ The activities	■ Overall completion date
■ The dependencies between the activities	■ Activity durations based on man/day resource inputs
■ Interim deliverables completion dates	

Tender strategy

There are two parts to the strategy: tender preparation and selling/profit recovery.

All contractors have to win projects at prices that produce a profitable outcome. Bidding strategy must recognise the need to decline invitations to tender for work likely to fail to meet the organisation's objectives, such as minimising risk, meeting a required profit target and failing to meet payment terms.

Unique problems and risks must be identified, planned and priced during the tender period. The bid management team must identify and analyse all such issues by reference to their own expertise, and experience, coupled with internal and external advice from experts, and specialty contractors.

Special solutions can be adopted for challenging projects:

- Just-in-time deliveries
- Off-site storage and prefabrication
- Stringent site rules and procedures covering restricted access, safety and security
- Early selection of experienced and reliable specialty contractors
- Special measures to limit and contain noise, dust and other sources of nuisance
- Site decontamination prior to commencement of general works
- Access solutions that provide safe working platforms above or beside site hazards
- 24/7 working and triple shifting with special health, safety and welfare facilities and transportation, to make working conditions as acceptable as possible for operatives facing particularly demanding requirements

- Restricted possession periods
- Restricted working periods.

Solutions must be researched and planned, including all necessary risk analyses, diagrams, drawings, flowcharts, critical path networks, work schedules and method statements, so that they can be priced and demonstrated to consultants and clients. The costs will be built into the preliminaries and/or works, and the written and drawn information will form part of the tender submission.

Compliant bids

What constitutes a compliant bid must be established:

- Selection criteria (e.g. a high level of quality in the finished project may require high levels of supervision)
- Method statement
- Project duration, sequence or timing
- Sectional completion and/or phasing requirements
- Security conditions for sensitive sites, in relation to either the site or its surroundings
- Provision of drawings of services as installed, for maintenance purposes
- Commissioning and testing
- Training of client's staff in operating services or mechanical equipment; some advice or recommendations may also be sought.

This detailed plan can be incorporated into a demonstration to the client on the value of employing a particular contractor.

2.4 Examine documents – contract, design, bonds, warranties and insurance

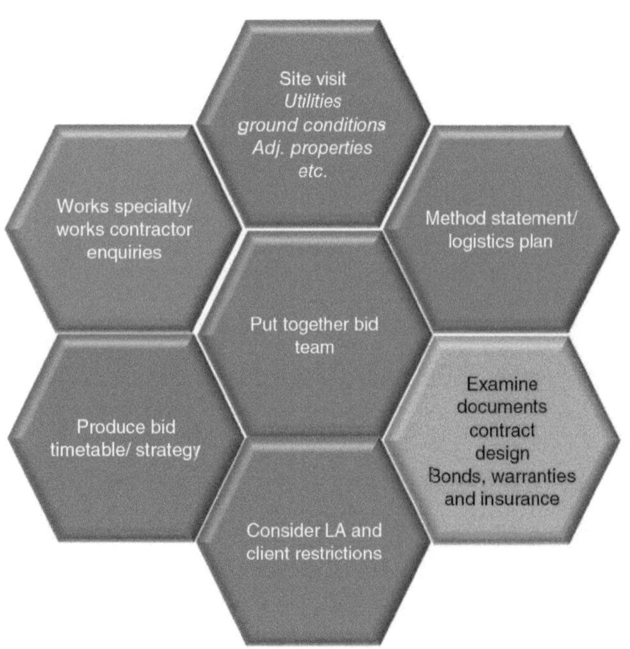

A check must be made to see that all drawings received are of the revision noted, and that all documents listed in the invitation letter are provided. A letter should be sent to the tendering authority to acknowledge receipt of the documents and confirm that a tender will be produced by the due date. This letter should also record any discrepancies in the documents received.

An accurate record of tender documents received must be kept; it will form the basis of a formal offer and checked against the contract documents in the event of a successful tender.

Information required

The information in the tender is supplied in various forms:

- Drawings
- Specifications (including performance specifications where appropriate)
- Schedules
- Technical reports
- Scheduled work periods, sequence and dependencies
- Employer's requirements
- Bills of quantities.

Tender submission documents vary considerably depending on the form of contract and documents used to define the works. Irrespective of the procurement type, the estimating principles are the same, whether it be:

- Lump sum based on drawings and specification
- Drawings and schedule of rates for a two stage tender
- Drawings and bill of quantities. Project can be awarded on the basis of a bill of approximate quantities, re-measured upon completion
- Design and build with a fixed lump sum price or target price where insufficient design detail/employer requirements are known at the time of tendering, based on single- or two-stage tender
- Drawings, specification and detailed cost plan
- Schedule of rates with re-measurement on completion
- Management contracting arrangement where the employer appoints a management contractor to oversee the works. Construction is completed under a series of separate works contracts, which the management contractor appoints and manages for a fee
- Construction management arrangement where the employer appoints separate trade contracts to carry out the works, and a construction manager to oversee the completion of the works for a fee, with target price or guaranteed maximum price
- Prime cost arrangement that requires an early start on-site, often for alterations or urgent repair work, such as flood damage. The exact nature and extent of the work is not known until the project is underway, so full design documents are not completed until work has commenced. The contractor is paid the prime cost plus a contract fee
- Public–private partnership where the special purpose vehicle submits a tender.

Checking validity

A checking procedure is needed to confirm that the project conforms to the information already provided. Drawing lists can be checked against drawing issue sheets that usually accompany the documents. It is not necessary to produce issue sheets for suppliers and specialty contractors because their documents are recorded on enquiry abstracts and listed in enquiry letters. A record of any subsequent additional or revised documents should be carefully recorded in the estimator's file.

Technical reports

Site investigation report, including water table and any exceptional items. This information is needed for piling, excavation methods, disposal of surplus excavated material and to consider the effect of hazardous materials and chemicals on permanent works.

2.5 Site visit – utilities, ground conditions, adjacent properties and so on

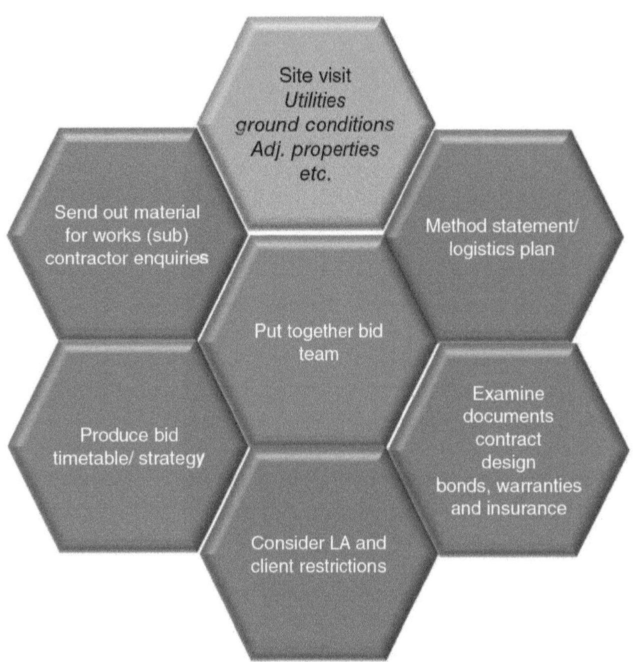

A site visit should be made as soon as a preliminary assessment of the project has been carried out and a provisional method and sequence of construction established. A comprehensive site visit report should be prepared with site photographs. A site visit video is particularly useful.

Solutions can be developed to address issues such as safe access, appropriate site facilities and environmental and considerate contractor measures required. The estimator will usually be assisted in this task by construction managers or a health and safety manager.

The proposed cost of the solutions will be included in the preliminaries and details passed to specialty contractors or used in order to prepare estimates for items such as scaffolding, site accommodation, waste removal and recycling. Valuable advice can be obtained from specialty contractors on alternative solutions and safe methods of working. Quotations received will form an important part of the preliminaries section of a tender, including the site establishment, access solutions, craneage and lifting.

Health, safety, welfare and environmental issues must be high on the agenda at tender review meetings. If a design is inherently unsafe or the client is not prepared to make due allowance for adequate site welfare facilities, it may be necessary to decline to tender or insist that the issues are properly taken into account by the client and their consultants. Under CDM (2015), the designer must eliminate health and safety risks to anyone affected by the project (Table 2.1).

Design review

The contractor will review the design for the project and divide it into suitable packages for suppliers, specialty contractors and consultants.

Visit to consultants

The estimator may need to visit the consultants, particularly when further information is needed which has not been given to tenderers, such as additional drawings and site investigation reports. Visits will normally be made to the architect, but it may be advantageous to visit the consulting engineer, services engineer and quantity surveyor, in order to meet the personalities who will be involved with the project.

Table 2.1 Points to be considered on a site visit.

Position of the site in relation to road and rail and other public transport facilities.	Ground conditions, any evidence of surface water or excavations indicating ground conditions and water table.
Site's topographical details including note of trees and site clearance required. The local authorities may be able to give advice on local conditions and any excavations which may be visible, adjacent/ near the site	Security problems, evidence of vandalism; hoardings, lighting, required
Hoardings and site access requirements, including any crossovers for pavements	Labour availability/skills shortages in the area. Visits should be made to local labour agencies and suppliers in the area and note made of other ongoing projects likely to impact labour availability
Names and addresses of local and statutory authorities	Location and availability of existing services, water, sewers, electricity, telephones, overhead cables, etc.
Special fire risks, fire brigade requirements	Security requirements in the area for the protection of equipment and materials
The effect any client requirements on access, storage, movement or site accommodation	Restraints imposed by adjacent buildings and services, i.e. space available for tower cranes, overhang, etc.
Facilities in the area for the disposal of waste and spoil	Location of nearest garages, hospital, police and cafés
Nature and use of adjacent buildings, such as industrial or residential	Other work currently in the area, or shortly to start
Possibility of using gantries and any footpath restrictions	Availability of space for site offices, canteen, stores, toilets and storage
Police regulations and parking restrictions for unloading of materials	Facilities for parking site vehicles and site workers' vehicles
Site access points and any restraints on layouts	Local transport availability for workers or need to supply transportation to site
Any demolition work, or temporary work needed to adjacent buildings and state of existing buildings with waterproofing requirements	Presence of hazardous materials (e.g. asbestos, contaminated ground) and any hazards to health. A CDM health and safety plan that identifies hazards likely to be encountered during construction, stating where and when they are likely to occur. The risk of a particular hazard occurring must be assessed. The safety plan (generated from the pre-construction Information) must be sufficiently developed to form part of the tender submission.

Detailed drawings, reports of site investigations and any other available information must be inspected and notes and sketches made of all matters affecting either construction method, temporary works or the likely cost of work.

Visits to consultants have become less common; contractors are usually given electronic copies of the drawings and specifications that become part of the contract documents. It is more likely that a visit will be seen as an opportunity to show that the tenderer is willing to contribute to the scheme, to work closely with the design team and to express an interest in further work. Identifying missing information and establishing a list of queries for the consultants should be undertaken so that letters can be written during the tender period or raised at a mid-tender interview. The opportunity to review the answers to questions raised by other contractors is also essential as issues might affect the risk assessment or cost of the work.

A critical assessment must be made of the degree of advancement and quality of the design, with respect to the RIBA design stages. A well-developed and well-documented design may be indicative of a smooth running and possibly profitable project. A design which is deficient and incomplete may delay progress, and the time or cost effect may not be at the risk of the employer.

Clearly defined contingency allowances should be considered in the tender. An adverse report may lead to the reconsideration of the decision to submit a tender.

2.6 Method statement/logistics plan

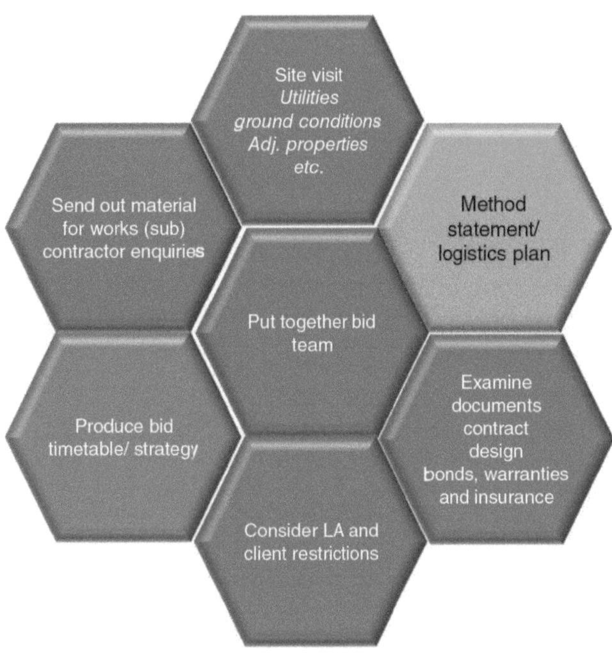

The method statement is an important document at the tender stage. It helps to establish the construction programme, plant requirements, plant location, the works package interfaces, location of the site accommodation and the factors influencing the sequence of operations. It will identify the temporary works requirements, such as temporary utilities and hoarding, access points and hazards that will/must be addressed.

A method statement is how the work is to be done, and a construction programme is when the work is to be done. The pre-tender method statement should be accompanied by a schedule, but, due to the limited time available in the tender process, this usually only consists of critical milestones and activities with long durations. The value/success of a method statement is based on the ability to identify and itemise work required. This can be done by using a work breakdown structure (WBS), which also informs the tender programme and, if the bid is successful, the production programme.

The method statement should include the possible dangers/risks associated with the relevant part of the project and detail the methods that will be established to control those risks and how the work will be managed safely.

Method statement

- ✓ Introduction
- ✓ Scope of works/objectives
- ✓ Location of site accommodation, batching plant hoardings
- ✓ Location and type of plant and equipment
- ✓ Sequence of working
- ✓ Hazards
- ✓ Material delivery and storage
- ✓ Methods for the removal or spoil
- ✓ Temporary supports
- ✓ Managing temporary utilities
- ✓ Traffic management
- ✓ Work areas for the work/trade packages
- ✓ Location of specialty contractors' equipment and storage facilities

The method statement can be submitted to the client with the tender, sometimes it is a requirement, and should include the identification of significant health and safety risks and how these will be controlled. This initial risk management process helps to satisfy the health and safety regulations at an early stage. A planning/costing statement may be produced by the pre-tender planning team to assess the plant, labour and method required for each of the major items in the Bill of quantities. This statement will include estimates of both prices and the timescales required. This can be a useful document, if the bid is successful, to provide the site management team with a record of how the tender was planned and priced. An important part of any of these statements is information on temporary works as these have a significant impact on costs and duration.

The site visit. This is especially important to understand the site traffic management and movement, delivery and logistics issues. Method statements from specialty contractors are co-ordinated to manage and understand the interfaces between the different specialty contractors. The method statement and the more detailed pre-construction statement should include enough information to ensure that quality standards (required by the client) are achieved.

The method statement should be included in the health and safety file where it can be examined in the case of an accident. Whilst generic method statements can be used as a basis to work from, site-specific method statements are vital and are compiled following a site investigation. They would include more detail about the work packages, including the resources needed as well as associated risk assessments.

A sample method statement is shown in Fig. 2.2. There are advantages in having a method statement for the client and the site labour force and management. Importantly, it outlines the risks involved and the safety measures that need to be employed. It gives those involved a greater understanding of the processes and resources involved and thus improves teamwork.

Reality

Reality must prevail. Time, cost, quality and safety and health are all major considerations in tendering. In a perfect world there is all the information available, plenty of resources and time is available to prepare the bid, but the reality of bidding often means that bids are prepared with a large risk contingency.

The method statement at the bid stage must be sufficiently detailed to understand how the project will be built. Once a contract has been awarded, more detail would be included at the pre-construction stage to produce a detailed method statement linked to the construction programme.

For small projects, the method statement will be brief, but nonetheless, very important. Producing a method statement for a refurbishment project adds complexity to the planning of the methods of production.

On large projects, the tender requirements may necessitate the contractor to submit a method statement.

A new dimension to submitting tenders – fly through and virtual reality

On mega projects and major infrastructure project proposals, contractors have produced electronic animated video presentations showing virtual construction, for submission with the bid. It explains to the design team the method and sequence of work, including photo realistic visualisations and 3D animation fly-throughs. A 3D fly-through animation, which is more commonly described as a fly-through, is a video that is created using 3D software showing what would be seen if the product had been created. Producing the document involves exporting the CAD files in the appropriate format into specialist software.

High-resolution 3D graphics and image rendering brings the tender documentation to life. Computer-generated images complement an otherwise text heavy visual

Company/Organisation			
Address			
Project/Task			
Location			
Project/Task ref.		MS No.	MS date
Scope of work Location, work limits, site access etc.			
Known hazards List hazards identified in the risk assessment and those hazards associated with project/task's proximity to other activities. Identify any hazardous substances to be used.			
Landlord/site owner arrangements Include access arrangements, any 'permits to work' etc.			
Responsible person(s)			
Briefing/communication arrangements			
Monitoring How will you monitor the project/tasks health & safety and environmental performance? Are workers aware of the incident reporting process?			

Figure 2.2 Method statement template – an example.

Operational Sequence How will the work be structured and organised to be carried out in a safe manner? How controls detailed in the risk assessment will be implemented. What PPE will be required?	
Permits/Authorisations Are any special permits required, e.g. hot works, confined spaces, statement of services etc.?	

Procedure for safe working	
Personnel safety arrangements	
Labour List competency & relevant qualifications of labour you are using	
Plant/Equipment Detail the equipment to be used explaining safe working practices, statutory checks and relevant operator training qualifications	

Figure 2.2 (*Continued*)

Materials Identify critical and long-lead delivery times materials to be used and potential issues such as manual handling, storage and disposal	
Deliveries Identify routes and drop locations	
Utilities Identify location of utilities and requirements for temporary water and power, including generators. Will any of the specialty contractors need special arrangements?	
Emergency arrangements Identify first aid & fire or other emergency procedures, first aiders and location of first aid and fire equipment.	
Environmental management Detail controls of harmful emissions to air, water & land.	

Name			
Position			
Signature		Date	

Figure 2.2 (*Continued*)

proposal. These images can also show construction processes in a simple but exciting manner and can form an integral part of a bid proposal.

High-quality 3D animation of the proposed methodology and project winning solutions can inspire and engage stakeholders.

A 4D simulation is a process that takes a 3D fly-through animation to higher level by accurately synchronising the build process to a project timeline.

Fully immersive virtual reality sequence, which means that with the aid of a low-cost VR headset, the client can immerse themselves into a virtual world, allowing them to see the construction process unfold as if they were actually on the building site watching construction.

Tender settlement

At tender settlement, the company directors will need to see that their legal responsibilities, as well as company policies, have been taken into account in the proposed method, resources and pricing.

Logistics **See Logistics in Principles section**

3 Contract terms and conditions

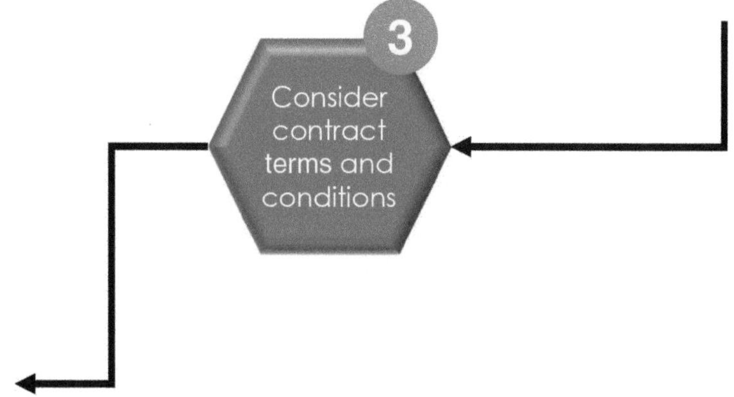

New Code of Estimating Practice, First Edition. The Chartered Institute of Building.

Many standard forms of contract are encountered in construction in the United Kingdom and overseas (see Table 3.1 for UK contracts):

- JCT 2016, with or without quantities alternatives, two stage, design and build, and management contracting alternatives
- Intermediate form tailored for medium-sized building projects, needing a less-complex contract
- the JCT Agreement for Minor Building Works used on smaller projects
- NEC4 suite of contracts used on many infrastructure projects. The New Engineering Contract was first published in 1993, with NEC4 in 2005 and NEC4 in 2017

Table 3.1 The contracts most commonly used in the UK.

NEC4		**FIDIC**
Engineering and Construction Contract (ECC)	Engineering and Construction Short Subcontract (ECSS)	Conditions of Contract For EPC/Turnkey Projects – The Silver Book
Engineering and Construction Contract – priced contract with activity schedule	Professional Service Contract (PSC)	Conditions of Contract for Plant and Design-Build – The Yellow Book
Engineering and Construction Contract – priced contract with bill of quantities	Professional Service Short Contract (PSSC)	Short Form of Contract – The Green Book
Engineering and Construction Contract – target contract with activity schedule	Term Service Contract (TSC)	Conditions of Contracts for Construction – The Red Book
Engineering and Construction Contract – target contract with bill of quantities	Term Service Short Contract (TSSC)	Conditions of Sub-Contract for Construction – for Building and Engineering Works Designed by the Employer
Engineering and Construction Contract – cost reimbursable contract	Supply Contract (SC)	Conditions of Contract for Design, Build and Operate Projects (DBO)
Engineering and Construction Contract – management contract	Supply Short Contract (SSC)	Construction Contract: Conditions of Contract for Construction MDB
Engineering and Construction Subcontract (ECS)	Framework Contract (FC)	
Engineering and Construction Short Contract (ECSC)	Adjudicator's Contract (AC)	
JCT		**CIOB**
Standard Building Contract	Construction Management Contract	Mini Form of Contract
Intermediate Building Contract	JCT-Constructing Excellence Contract	Mini Form of Contract for Home Improvement Agencies
Minor works Building Contract	Measured Term Contract	Small Works Contract
Major Project Construction Contract	Prime Cost Building Contract	
Design and Build Contract	Repair and Maintenance Contract	Concise Building Contract
Management Building Contract	Home Owner Contracts	Domestic Building Contract

- FIDIC forms of agreement. FIDIC is a French language acronym for Fédération Internationale Des Ingénieurs-Conseils, which means the International Federation of Consulting Engineers. FIDIC is headquartered in Switzerland; it produces standard forms of contract for civil engineering works that are used internationally. The contract is often referred to by the colour of the cover:

  Minor works – Green book

  Employer design (traditional projects) – Red book

  Employer design with multilateral development bank providing finance – Pink book

  Contractor design (traditional project) – Yellow book

  Engineer procure construct, Turnkey project – Silver book

  Design, build and operate project – Gold book.

3.1 Special employer requirements and modifications to standard clauses

Alterations to conditions and terms in standard forms of contract can have a significant effect on the contractor's liabilities, the risk allocation and the roles and responsibilities of the parties to the contract. Such changes must be highlighted by the estimator (or contract specialist) and the cost implications incurred in meeting the new responsibilities included in the bid. Typically, such amendments may include the requirement that the contractor is responsible for ascertaining all ground conditions and no claim will be entertained for adverse ground conditions not taken into account in the tender submission. This means that all ground conditions are at the contractor's risk.

Supplementary conditions are often attached to contracts, dealing with insurances, warranties, collateral warranties, design responsibilities and performance bonds.

Bills and specifications may require the contractor to carry out design work where the contract conditions contain no design provisions. Where onerous non-standard conditions are attached to a contract, contractors may either ask to be excused from tendering or add a premium to the tender bid.

4 Resource planning and pricing

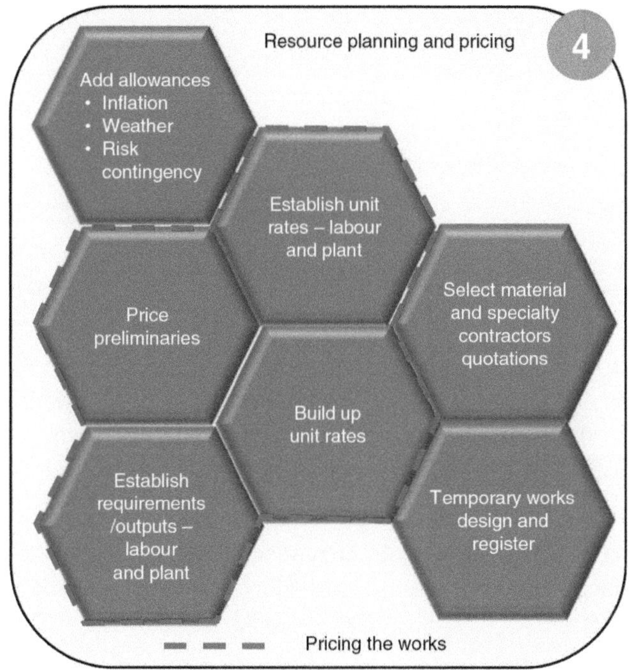

4.1 Pricing the works

Pricing a project based upon drawings and specifications, a bill of quantities (BoQs) or any other method of procurement embodies the same principles, cost, price and value. The cost is the production cost, the price is the cost plus a margin for profit and the value is what the customer believes the item to be worth to them. Figure 4.1 shows the cost build-up.

- Direct costs are the materials, labour, plant and equipment, temporary works and all directly involved efforts or expenses in the production of the item.
- The variable and ***indirect*** costs are those not directly related to production, although the business could not function without them. They include costs which are frequently referred to as overhead expenses (e.g. rent, insurances and information and communication technologies) and general and administrative expenses (e.g. director's and manager's salaries, estimating department, marketing department, purchasing and procurement department, planning department, accounting department costs, human resources department costs and equipment depreciation).
- Some overhead costs are attributed to the project and treated as direct costs. Site overheads are calculated and priced in the preliminaries and other items. Head office administrative overheads (back office costs) are calculated as a percentage of annual revenue/turnover with recovery included in the price build up as a percentage of the project price.
- The gross profit is a percentage of the overall direct and indirect costs. Profit will vary from project to project, dependent upon the risk and complexity of the project.
- The risk contingency/allowance reflects the risk associated with the production of the item and the project. Within the risk items will be allowances for price inflation over the duration of the project, the risk of failure to complete on time and the possibility of paying liquidated and ascertained damages. A large liquidated and ascertained damages amount will lead to a risk contingency being added.

The contractor deals with resources (human, materials, plant, equipment, financial, time), the tasks to be undertaken, the measured quantity, the quality required, the method of working (buildability), sequence of work, location of the work items, interrelationship with other items that could cause disruption and delay, any temporary works required to undertake the task and the duration.

When bidding, ultimately the contractor, specialty contractors and suppliers will make a decision about the market and the likelihood of winning work at a particular profit margin.

Prices are broken down into their constituent parts (labour, materials, plant and overheads). On large projects, the estimator is reliant upon the bid team to provide detailed information about the sequence, method of production, required duration, plant requirements and logistics. On smaller projects, the person undertaking the work on the job site may prepare the estimate and consider every aspect of the project from procurement to production.

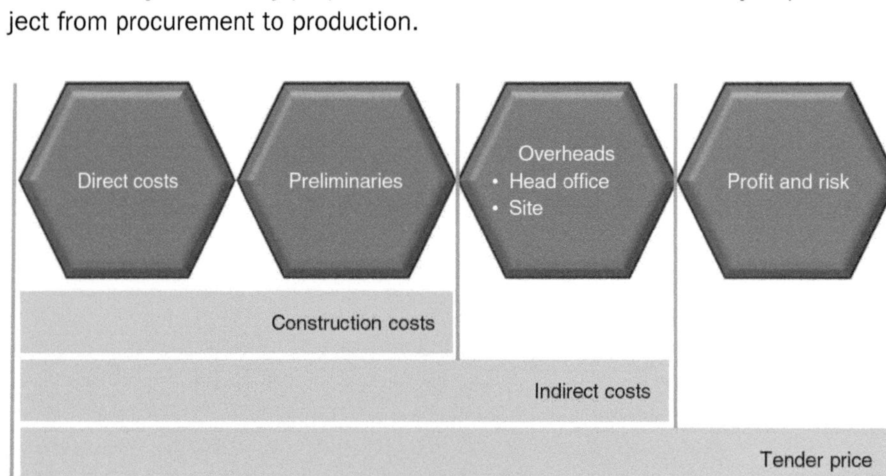

Figure 4.1 The cost build-up.

Pricing building work uses the analytical or 'bottom up' estimating process, by building up unit price rates for defined items, pricing of site based and head office project overheads and dealing with prime cost (PC) and defined and undefined provisional sums. Figure 4.2 shows the items to be included in the build-up of price rates, showing the items considered in a price build-up. The honeycomb comprises six main parts – labour, materials, plant, allowances, wastage and overheads and profit.

Computer systems

Computer systems are fundamental to the pricing of projects. Measurement can be taken from the CAD drawings or using a 2D system that can measure from scanned pdf or CAD information. Some software packages have full 3D graphics capability and have the ability to extract object properties and generate automatic quantities from 3D BIM information.

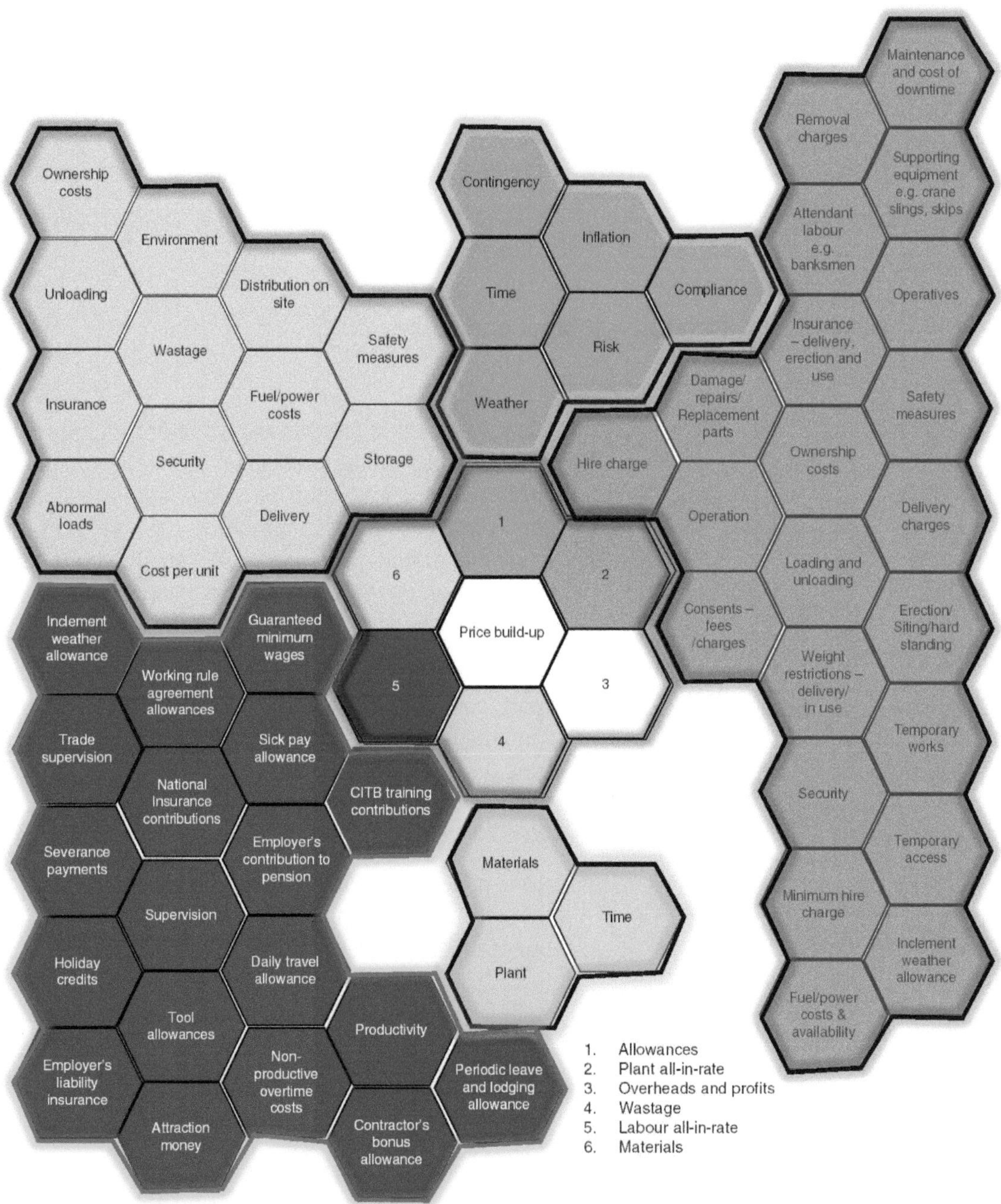

Figure 4.2 The pricing honeycomb.

Spreadsheet-based systems can manipulate the documents for measured items into cost plans and specialty trade/work packages. Previous documents can be imported and used as a template, access rate libraries, enter codes for sorting and generate reports. Some companies have their own database of prices for measured items, while others will build up prices for each of the items using standard estimating software that embodies a comprehensive cost library. The software provides the mechanism for developing the bid. The estimator's analysis of costs is made using a computer model of a bill, which is entered in a number of ways:

- Bills of quantities are available electronically for immediate incorporation into a computer-aided estimating system
- Scanning written bills of quantities to a text file
- Importing the CAD files for electronic measurement.

Software used in bidding and tendering provides a management system that tracks progress and helps keep information up to date, accessible and secure. Reports can be produced easily, and all current tenders can be managed through one interface. The software supports good communication, enabling the distribution of files and messages to the bid team and supply chain. Less time can be spent on each tender, thus reducing administration costs. The management of electronic documents throughout the process reduces printing (and distribution) costs.

Bid/tender software incorporates a secure system for maintaining confidentiality, particularly if the information is cloud based.

4.2 Establish unit rates – labour and plant

Principles

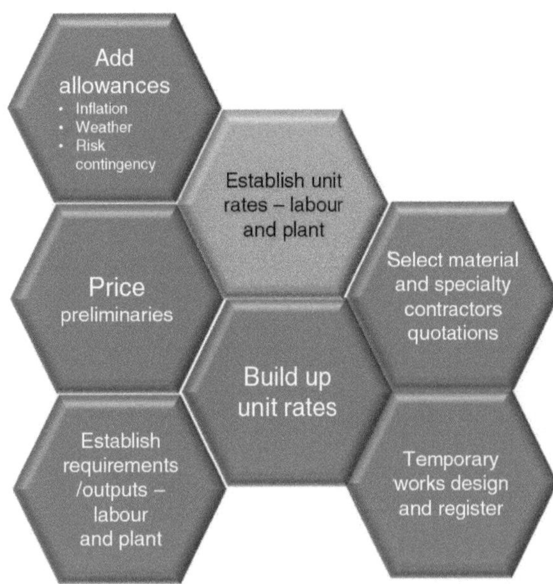

Consideration is given to every factor that may influence the cost of the work when calculating unit price rates. Where work has been measured in accordance with a standard method of measurement, the definitions and coverage rules should be clearly understood, as these affect the pricing.

For example, the measurement of a drainage trench is deemed to include any necessary earthwork support, compaction, backfilling with excavated material and disposal of surplus soil. This would not be obvious just from reading the description of the item of work for the trench, but a contractor should make due allowance in the unit price rate.

Unit price rates for measured items in the bills of quantities (excluding preliminaries) consist of any or all of the basic elements:

- *Labour*, comprising the gang rate for both operative skilled and unskilled labour, the allowance for productivity with any allowance built in for unproductive time, which includes waiting time for other trades; bearing in mind, the labour rate will not distinguish between small and large areas and complex activities. Wage rates in the Working Rule Agreements (WRA), whilst providing the benchmark for wage rates, operatives in certain regions will not work for the agreed minimum wage rate, they charge a premium for their hourly or daily rate because of scarcity of skills.

Advantages of a computer-aided system

- Faster turnaround of tender documents
- Bid team integration Improved collaboration
- Direct take-off from drawings
- Better accuracy with integration between takeoff figures and the production of an estimate
- More efficient tender adjudication
- Improved risk management
- Contract information maintained in a single location
- Improved customer communication
- Store documents in one location for easy access
- Access historical cost information
- View specialty contractors' pre-qualification details
- Track supply chain

- *Materials*, delivered to site in full or part loads, including transport costs and trade and quantity discounts, with allowances for wastage of materials, storage and warehousing requirements, handling, double handling, supply flexibility, delivery frequency, pricing flexibility and minimum load size.
- *Plant, special plant, equipment and tools* required for the specific item, general static plant will be included in the preliminaries section, including water supply, electricity, petrol/diesel, grease and waste disposal. Allowances must be made for transportation, loading and unloading, insurances, breakdown, down time and cleaning. Small tools and equipment should be allowed dependent upon the tasks.
- *Specialty contractors* undertaking work/trade packages, together with any attendance allowance required for the works package to be undertaken. Some works packages may include an element of design that will need to be priced.
- *Overheads* (site based and head office).
- *Profit/mark up* is defined as a return on investment. This is normally judged or assessed rather than calculated. The margin depends on the judgement of the market and the state of the business. It is what that provides the contractor with an incentive to perform the work as efficiently as possible. The addition for overheads and profit is often part of the tender settlement process discussed later.

Variability, accuracy and consistency of prices

There will be variability of prices between contractors for the same item of work. The reasons for variability arise from the price being an estimate of the cost to carry out the work. Production systems will differ, productivity rates will differ and there will be different views on the amount to be included for lost time caused by weather, absenteeism and interruption of the workflow by change orders. Different estimators may interpret the same item in a different way. Judgement, knowledge and experience will all influence the pricing process.

Innovation will also play a role, with contractors devising different approaches to the same item of work resulting in price differences. Accuracy and consistency are important requirements for any contractor pricing work. Accuracy implies a closeness to the actual value, while consistency is ensuring that the pricing follows the same path in relation to its accuracy. Any contractor submitting inaccurate bids will quickly need to adjust the pricing to reflect the real costs following feedback that the project has not delivered the anticipated returns.

Pricing strategy

There must be a clear strategy for pricing work items. It is unlikely that items will be priced in the order they appear in the bills of quantities, because a better understanding of activities is gained by pricing one work section at a time. Computer estimating systems sort the bill items into work section order (similar items) or any order required. Computers allow resources to be entered either through a resource build-up screen for each item or with the aid of a spreadsheet-type comparison system, where similar trade items can be viewed in a single table.

Chasing revenue (turnover) growth in business is vanity; Chasing profitability is reality.

If quotations for materials are delayed, the estimator can price labour and plant first and return to part-priced items later when quotations are available. On the other hand, 'typical' materials prices may be used during the pricing stage using prices from similar projects.

Major contractors will have bulk supply prices, sometimes called blanket orders, spot contracts or blanket discount contracts, for procuring key materials on a bulk supply contract. Better economies of scale are achieved by aggregating orders over a time horizon.

4.3 Establish unit rates – labour, materials and plant

The contractor needs to know:

- How the work will be undertaken
- The sequence and duration of the work
- The availability of people, resources and materials
- Any special risks involved.

Pricing alterations work is more complicated because of the large number of unknowns.

A net unit rate for an item of work is built up in four distinct stages:

1. Establishment of all-in rates for the key items that will be incorporated
2. Selection of methods and production standards
3. Incorporation of price rates from specialty contractors' quotations with appropriate adjustments for attendance, support and profit
4. Calculation of overheads (site-based direct costs and head office indirect costs).

Stage 1: Establishment of 'all-in rates' for the key items that will be incorporated

Labour: A rate per hour for the employment of skilled and general operatives. Wage rates are established for the different categories of labour used on the project. Labour must make allowances for overtime working, sickness pay, holidays with pay, non-productive time, tool allowances and allowances for travel time to and from the site in accordance with the Working Rule Agreement (WRA).

Materials: A cost per unit of material delivered and unloaded at the site, including transportation and any unloading and special plant required for handling. Comparison of the various quotations received for materials and the selection of one of these for use in the bid. Assumptions will need to be made about price inflation and how long the quotation is open for acceptance. Until a project has been awarded and the contract signed, quotations accepted or orders placed.

Storage, temporary protection of special materials, may be included in the Preliminaries section. Allowances for wastage must be included.

Off-site pre-fabrication of components is increasingly used, such as pre-assembled plumbing systems, pre-cast concrete structural components and bathroom pods. The pre-fabricated units must be delivered and stored with due allowance made for handing and fixing. The cost of logistics is an important part of pricing such components.

Plant and equipment: The operating rates from plant hire organisations per hour (or per day, per week, etc.) for items of plant and equipment, including fuel, cleaning and transportation to and from site. The plant is supplied with or without the operator and the insurance requirements. Alternatively, the rates from the contractor's own data/ plant department are used.

Stage 2: Selection of methods and production standards

The selection of methods and production standards/productivity outputs from the contractor's database or other sources. The standards are used with the all-in rates calculated in Stage 1, to calculate net unit rates, which are set against the items in the bills of quantities.

Alternatively, unit price rates received from specialty contractors for works packages are used, with a suitable addition for management, attendance and profit.

Stage 3: Incorporation of price rates from specialty contractors' quotations with appropriate adjustments for attendance, support and profit

The incorporation of unit price rates from specialty contractors, including those offering labour-only services, producing either the whole or part of a rate. Incorporation of any special allowances required, such as insurances, special licenses and temporary works considerations. In some situations, the specialty contractor will give a lump sum fixed price for the work without a unit price rate breakdown, in which case the estimator must allocate prices to the measured items.

Stage 4: Calculation of overheads (site-based direct costs and head office indirect costs)

The calculation and addition of site-based and head office project overheads is a separate and subsequent operation in the pricing of the preliminaries. The contingency allowance for risk added at the tender adjudication stage.

Working Rule Agreements

There are a number of national WRA in the United Kingdom for construction and allied trades; similar agreements exist overseas. The Construction Industry Joint Council's (CIJC)* WRA for the Construction Industry covers over 500,000 workers. It is the largest industrial agreement within the UK construction industry. There are pay rates for different skill bands, craft and general operatives. The employers' representatives and the union representatives regularly review the WRA.

The Building and Allied Trades Joint Industrial Council's (BATJIC)† WRA is for members allied to the Federation of Master Builders; the principles are similar to the CIJC agreement. The Joint Industry Board for Plumbing and Mechanical Engineering Services (JIBPMES) negotiates wage agreements for the specialist services. They have an agreed 37.5-hour week.

Since June 2018, the basic hourly rate payable to the operative for a technical plumber and gas service technician, inclusive of tool allowance, is £16.73, for an advanced plumber and gas service engineer £15.07 and for a trained plumber and gas service fitter £12.93. Apprentices have 4 years of training, reaching NVQ level 3.

The Joint Industry Board for the Electrical Industry‡ negotiates wage agreement for electricians, electrical/site technician, mechanical technician and cable installation supervisor (or equivalent specialist grade), with own transport from January 2018; the agreed basic hourly rate is £18.37, and in London 20.57. Both CIJC and BATJIC are used to explain the build-up of the labour rates in this code of estimating practice (CoEP).

There are six basic rates of pay and rates for apprentices, a general operative rate, four rates for skilled operatives, a rate for a craft operative and rates for apprentices. The WRA lists the skill categories.

* Construction Industry Joint Council, Working Rule Agreement for the Construction industry, 2013 with revised wage rates published annually. Construction industry Publications Ltd, London. 79p ISBN : 9781852631345.

† Building and Allied Trades Joint Industrial Council, Working Rule Agreement 2013/2014 with updates on wage rates published annually. Building and Allied Trades Joint Industrial Council, London. 29p

‡ The Joint Industrial Board comprises The Electrical Contractors' Association and Unite the Union.

Pay rates – CIJC (valid from June 2017)

	Per hour (£)	**Per week (guaranteed minimum weekly earnings, £)**
Craft rate	11.93	465.27
Skill rate 1	11.36	443.04
Skill rate 2	10.94	426.66
Skill rate 3	10.24	399.36
Skill rate 4	9.67	377.13
General operative	8.97	349.83

Note: Skill rates are the specialisms coded under the CIJC. For example, gangers and chargehands are in skill rate 4, formwork carpenters in the craft rate, apprentice year 1 trainee skill rate 4, and apprentice year 2 skill rate 3.

The WRA covers the following areas:

Entitlement to basic rates of pay: WRA defines general operative, craft operative and four skill rates. Rates of pay are reviewed periodically by the CIJC. Apprentices are entitled to be paid during normal working hours to attend approved courses for off-the-job training in accordance with the requirement of their apprenticeship. Payment during such attendance shall be at their normal rate of pay.

Working hours for construction, guaranteed minimum weekly earnings: The normal working hours are Monday to Thursday, 8 h/day; Friday 7 h/day and total 39 h/week. Breaks average 1 h/day.

Overtime is calculated as: Monday to Friday: For the first four hours, after completion of the normal working hours of the day at the rate of time and a half; thereafter, at the rate of double time until starting time the following day. Saturday: At the rate of time and a half, until completion of the first four hours, and thereafter at double time. Sunday: At the rate of double time, until starting time on Monday morning.

An operative, who has been available for work for the week whether or not working due to shortage of work, or delayed by inclement weather, shall be entitled to the guaranteed minimum weekly earnings based on the normal contractual working hours and guaranteed minimum hourly rate of pay.

Bonus, overtime rates: It shall be open to employers and operatives on any job to agree a bonus scheme based on measured output and productivity for any operation or operations on that particular job.

Periodic leave, annual holidays and public holidays: The holiday year runs from the second Monday in January each year. Operatives are entitled to 29 days paid annual holidays inclusive of eight public and bank holidays. Paid holiday entitlement accrues at the rate of 0.558 days/week of service. This is an absolute entitlement, which cannot be replaced by rolling it up into basic pay, bonus or any other allowance, which would result in the operative not receiving their full holiday pay when taking annual leave. The entitlement to statutory paid holidays continues to accrue during the entire period of employment, notwithstanding that the operative may be absent due to sickness, paternity/maternity leave and so on. BATJIC recommends that holiday pay should be based on 12.6% of weekly pay.

Daily fare and travel allowances: Operatives are entitled to a daily fare and travel allowance, measured one way from their home to the job/site. The tax authorities determine the tax-free amount that can be paid without incurring income tax.

Shift working, night work and continuous working: Where work is carried out at night by a separate gang of operatives from those working during the daytime, operatives shall be paid at their normal hourly rate plus an allowance of 25% of normal hourly rate.

Health safety and welfare: Employers and trade unions are committed to improving the industry's safety record. Major contractors are insisting that all those employed on their sites hold a current Construction Skills Certification Scheme CSCS[§] registration

§ CSCS is the skills certification scheme within the UK construction industry. CSCS cards provide proof that individuals working on construction sites have the required training and qualifications for the type of work they carry out.

card. To be issued with such a card, the operative must have passed a safety awareness test within 2 years of applying or re-applying for a CSCS card. The CSCS card remains valid for 5 years, and must be renewed by passing a further safety awareness test. At least one first aider must be present when the number of employees on-site, job or shop is between 50 and 150; there should be a least one additional first aider for every 150 or more employees. Where there are fewer than 50 employees at work, there may be no need for a first aider, but in this case, the employer must ensure that there is an 'appointed person' present at all times when employees are at work. Everyone working on-site will go through a health and safety induction process before they are allowed to commence work on-site.

Payment of industry sick pay, benefits schemes: An operative who, during employment with an employer, is absent from work on account of sickness or injury shall, subject to satisfying all the conditions, be paid the appropriate proportion of a weekly amount specified for each qualifying day of incapacity for work. For this purpose, the appropriate proportion due for a day shall be the weekly rate divided by the number of qualifying days.

Subsistence allowance: Operatives should be willing to travel a reasonable distance from their home and/or the main yard of the employer to carry out duties, but where an employer requires an operative to be available to lodge overnight on a permanent availability basis, then a mutual agreement should be established between the employer and the operative. An operative necessarily living away from the place in which they normally reside shall be entitled to a subsistence allowance.

Storage of tools: When practicable and reasonable on a site, job or in a workshop, the employer shall provide an adequate lock-up or lock-up boxes, where the operative's tools can be securely stored.

Abnormal conditions of work; Where an operative is required to work for 1 h or more in conditions of dirt, inconvenience or discomfort to an extent abnormal to the particular craft or trade, then they shall qualify for a conditions payment, to be agreed on-site.

Availability for work: An operative has satisfied the requirements to remain available for work during normal working hours by complying with the following conditions:

- Unless otherwise instructed by the employer, the operative has reported for work at the starting time and location prescribed by the employer and has remained available for work during normal working hours.
- Carries out satisfactorily the work for which the operative was engaged or suitable alternative work if instructed by the employer.
- Complies with the instructions of the employer as to when, during normal working hours, work is to be carried out, interrupted or resumed.

Where work is temporarily stopped, or is not provided by the employer, the operative may be temporarily laid off. The operative shall be paid the normal rate of pay for the day on which they are notified of the lay-off and one-fifth of the Guaranteed Minimum Weekly Earnings for each day.

Night work: An 8-hour limit per day for night work.

Accident and death benefit: Death benefit of £25,000 and provided on a 24/7 basis with the cover doubled to £50,000 if death occurs either at the place of work or travelling to or from work.

Protective clothing: Suitable protective clothing must be provided to operatives who are required to work in inclement weather such as rain, snow, sleet and hail or as required by government regulations.

Termination of employment, disciplinary procedure, trade unions

Industry Training Board-Construction Industry Training Board: As an Industry Training Board, the Construction Industry Training Board (CITB) are given powers to collect annual levy payments, with the money used to provide training grants and other services that support the UK construction industry. The levy and grant system is a collective solution that supports skills and training in the construction industry. The levy has been in existence in the United Kingdom since the 1960s.

The requirement to pay is dependent upon the wage bill. The total wage bill consists of the total pay-as-you-earn (PAYE) for a year and the total payments made to labour-only sub-contractors for a year.

If the total wage bill is £79,999 or less, companies are not required to pay the levy, but they do still have to complete a levy return. If the total wage bill is £80,000 or more, the levy must be paid. Companies pay 0.5% on the total PAYE for a year and 1.5% on the labour-only sub-contractor payments for a year.

Stakeholder Pension: The EasyBuild stakeholder pension was developed to address the issues of the construction industry, providing a flexible pension, which is suitable for all employees. Both the employee and the employer contribute to the pension scheme.

Apprenticeship Levy 2017

The UK government is committed to boosting productivity by investing in human capital. As part of this, the government is committed to developing vocational skills, and to increasing the quantity and quality of apprenticeships. The Government has created an employer-led model for developing and funding apprenticeship standards, replacing the current system and focusing on employers driving the development and delivery of apprenticeships. The levy will help to deliver new apprenticeships and support quality training by putting employers at the centre of the system.

All apprenticeships include elements of on-the-job and off-the-job training, leading to industry-recognised standards or qualifications. Some apprenticeships also require an assessment at the end of the programme to assess their ability and competence in their job role.

As at May 2017, the apprenticeship levy requires all employers to pay 0.5% of any wage bill over £3 million into the government's new apprenticeship service, to pay for apprenticeship course fees. There will be a levy allowance of £15,000 per year to offset against the levy that has to be paid.

For the purposes of the levy, an 'employer' is someone who is a secondary contributor, with liability to pay Class 1 National Insurance Contributions for their employees. The levy is payable through pay as you earn (PAYE) and will be payable alongside income tax and National Insurance. To keep the process as simple as possible, the 'paybill' will be based on total employee earnings subject to Class 1 secondary NICs.

Employers paying the apprenticeship levy will be able to access the funds they have paid for the cost of apprentice training through a new digital account. The government tops up this amount by an additional 10%.

The apprenticeship levy must be calculated in the overheads section to make due allowance for the overhead costs incurred in administering and paying the levy.

4.4 Labour cost issues – a summary

Labour costs	
Guaranteed minimum wages	Basic rates for each skill class of Operative. No distinction between male and female operatives
Bonus allowance	Bonus to retain and reward operatives
Inclement weather allowance and interruption of work	Normally included by paying full weekly wage rate in labour and including an allowance for inclement weather in the Preliminaries due to inclement weather interrupting the normal work pattern. Distinction is made between inclement weather interrupting the normal workflow and exceptionally inclement weather, which may be eligible for an extension of time to the project duration
Non-productive overtime costs	Rules for calculating overtime rates, some may be nonproductive, such as cleaning plant and equipment at the end of the work day to ensure smooth operations the following day
Sick pay allowance	Rules on the amount to be paid for sick pay and the length of time allowed are included in Working Rule Agreement
Trade supervision	Proportion of non-productive time by supervisors often considered in Preliminaries
Payments for discomfort, inconvenience and risk	Extra payments stipulated in the Working Rule Agreement for particular activities. Allowance should be made in the buildup of the all-in wage rate
CITB training contributions	Normally levied at 0.25% of entire payroll, and 2% for self-employed and labour-only subcontractors
National Insurance contributions	Employer contributions are a percentage of weekly earnings
Holidays with pay	Annual and public holidays with 30 days leave
Tool allowances	Consolidated within the basic wage rate, employer to be responsible for storage of tools. The operative will be responsible for provision of small tools
Redundancy and severance payments	Statutory scheme for redundancy payment. A severance payment is the compensation that an employer provides to an employee whose job has been eliminated or by mutual agreement has decided to leave the company, or who has parted ways with the company for other reasons. Typically, severance pay amounts to a week or two of pay for every year that the employee was with the company
Health and safety and first aid on the job site	Normally priced within the Preliminaries section as costs associated with health, safety and wellbeing. Costs of induction training should be included within the health, safety and wellbeing costs
Employers liability insurance	Considered in project overheads
Daily travel allowance	Scale of allowances for daily travel one way. Some employers provide transport, which reduces the allowances
Subsistence allowance	Subsistence and lodging allowances are given subject to tax rules
Supervision on the job site	Priced using tender works programme, generally within the Preliminaries section
Protective clothing	Normally included as an amount in the Preliminaries
Attraction money beyond the agreed wage rate	Needed in remote areas or sites which might experience shortages of local labour

Making allowances in pricing

It may not be possible to determine all the above factors with accuracy at an early stage of estimating. There are benefits in separating some of the items from the calculation of all-in rates with preliminaries costs and project overheads. Where the allowances and prices for the items are included it is usually a matter of opinion and company policy. The distinction drawn between items set out in all-in rates, preliminaries and project overheads should not be regarded as mandatory.

The important consideration is that due allowance must be made for the items to establish the true costs to incorporate adequate allowances made in the estimate. Nobody can control inclement weather; it is the impact on production that matters. The impact can be minimised by careful planning and measures taken to protect the project from disruption. Hence, allowances are just that, a contingency included to ensure the direct costs are covered.

4.5 Build-up unit rates

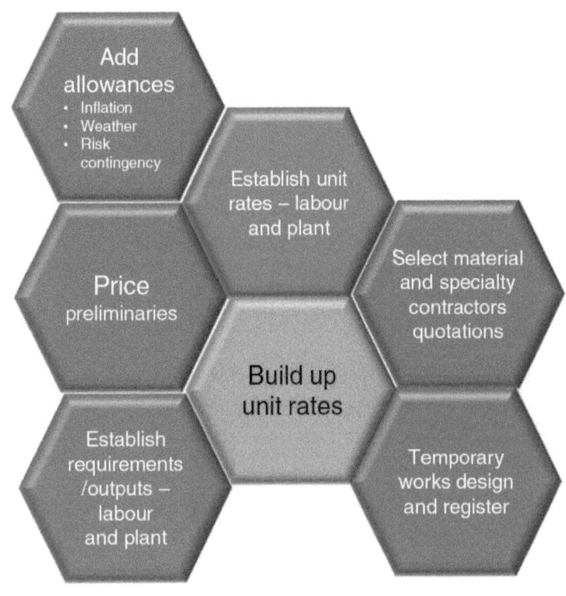

There are three steps in the calculation of an all-in unit price rate for labour:

Step 1: Determine the number of working hours an operative is expected to work during a 1-year period, taking account of the holiday arrangements in the WRA.

Step 2: Calculate the cost for the 1-year period for wages and the allowance for each item from the WRA items listed. Make an allowance for the impact of inclement weather on production time. Remember, this is not exceptionally inclement weather that is beyond what can normally be expected.

Step 3: Summarise the individual costs, calculate the all-in rate per hour by dividing the total costs by the total number of hours. Make allowances for the gang sizes for each of the work items.

For a very large project, it may be desirable to make a special calculation based on the anticipated construction period. Details vary according to:

- The trade of the operative (whether craft operative, general operative or skill rate)
- The firm
- The regional location of the project and the impact of skills shortages on the rate to be paid
- The industrial and legal conditions in force at any time.

The following calculations are based on a 1-year employment period and could apply to all projects for which tenders are to be submitted. It is for a craft operative.

Amendments must be made each time there is a variation in the cost of one of the factors included in the calculations or when further factors are introduced. Alterations must be made as soon as variations are promulgated, although there may be a period before they come into effect. The whole calculation should be revised regularly.

Step 1: Determination of hours worked

Summer period	**Hours**	**Winter period**	**Hours**
Starting time 08.00 Finishing time 17.30 (Friday 16.30) Lunch period 13.00–13.30pm Working week=44 hours for 30 weeks	1320	Starting time 08.00 Finishing time 16.30 (Friday 15.30) Lunch period 13.00–13.30 Working week=39 hours for 22 weeks	858
Deduct: 10 days summer holiday	(88)	Deduct: 9 days Christmas holiday	(80)
4 days Easter holiday	(35)	3 days public holiday	(23)
5 days public holiday	(44)		
Total hours for summer	1153	Total hours for winter	755

The number of hours worked during the calendar year, that is January to December, will depend on the hours worked per week during the summer and winter periods, with adjustments for annual and public holidays. The hours worked will vary between different companies, some variation in hours can be expected between firms operating in the north and south of the United Kingdom, due to the amount of natural daylight hours in the winter period. Local customs and availability of labour also affect the number of hours worked.

A company may agree to work hours suitable for a particular type of contract, due to special requirements of the employer or by special agreement with its employees. For example, on a refurbishment project, where work is being undertaken in the dry, it would be possible to work longer hours with artificial lighting in the winter.

Step 2: Calculation of hours worked

Gross hours available for work 1153+755	1908
Deduct allowance for sickness say 10 days	(88)
Net hours available for work (basic hours)	1820

Step 3: All-in hourly wage rate

Table 4.1 shows how the all-in hourly rates are calculated based upon CIJC WRA. Productive time has been based on a total of 1784 h (45.7 weeks) worked per year.

Overtime allowances

The overtime allowance is calculated as follows:

*Summer period***:**

Working weeks in summer 30 less 4 weeks for annual and public holidays = 26.

Non-productive overtime for summer period = 26 weeks × 2.5 h = 65 h.

	Mon	**Tues**	**Weds**	**Thurs**	**Fri**	**Sat**	**Total**
Hours worked	9	9	9	9	8		44
Basic hours	8	8	8	8	7		39
Overtime hours	1	1	1	1	1		5
Non-productive hours	0.5	0.5	0.5	0.5	0.5		2.5

Table 4.1 Rates per hour and per week – CIJC rates valid from July 2017.

		Wages @ standard basic rate as per CIJC WRA			
		Craft Operatives		General Operatives	
	Weeks	Week (£)	Year (£)	Week (£)	Year (£)
Productive time	45.7	465.27	21,262.84	349.83	15,987.23
Lost time allowance, say	0.85	465.27	395.48	349.83	297.36
Overtime allowance and enhanced payments such as bonuses, say	0.5	441.87	232.64	349.83	174.92
			21,890.95		**16,459.50**
Extra payments	45.7				
CITB levy (0.5% of payroll)	52		109.45		82.30
Holiday pay and Public holiday pay	5.8	452.79	2,698.57	349.83	2,029.01
Employer's contribution to:					
Easy Build Stakeholder Pension	52	7.50	390.00	7.50	390.00
National insurance (average weekly payment)	52	35.50	1,846.00	23.20	1,206.40
			26,934.97		**20,167.21**
Severance pay, sick pay, and sundry costs	1.5%		385.40	1.50%	287.78
			27,320.37		**20,454.99**
Employer's Liability and Third Party Insurance	2.20%		546.41	2.00%	409.10
Total cost per annum			**27,866.78**		**20,864.09**
Total cost per hour			**15.64**		**11.71**

Winter period:

Total for whole year is 65 h. Therefore, the cost of non-productive overtime: (basic rate per 39-hour week) = £178.62/39 = £4.58. Therefore, overtime allowance = £4.58 × 65 = £297.70.

	Mon	Tues	Weds	Thurs	Fri	Sat	Total
Hours worked	8	8	8	8	7		39
Basic hours	8	8	8	8	7		39
Overtime hours	0	0	0	0	0		0
Non-productive hours	0	0	0	0	0		0

Inclement weather

The time lost for inclement weather varies with the type of work, season of the year and geographical area. An allowance is used and any adjustment necessary made for exceptional situations in the project overheads.

Say the time lost due to inclement weather is 2%, that is ~36 h. This will vary from year to year, dependent upon the weather pattern and the geographic site location.

Actual hours worked = Basic hours less inclement weather 1820 − 36 = 1784

4.6 Gang sizes for activities

Each activity on the job site will require collaborative working in teams and gangs. Estimators will use their own assumptions for the appropriate gang size for each of the activities in accordance with their experience. The assumption is that the ganger, supervisor or foreman will spend a proportion, probably 40–60% of their time, managing, supervising, arranging materials and ensuring the safety and work flow. This is factored into the cost of the gang when the hourly cost is being considered. Table 4.2 shows the gang size cost, production rates and so on.

Production standards and other considerations

The records of cost and outputs achieved on similar work from previous projects are a major source of information used in estimating. This information arises from records of resources used on site or from work-study exercises to establish standard production rates. The production rates can be compared with the productivity factors in published price books and those in computer-assisted cost databases. Always use caution when using unchecked price data, there is no substitution for in-house figures and experience. Where work is to be awarded to a specialty contractor under a domestic contract agreement with the contractor, the specialty contractor will use the same pricing procedure to price their work schedule provided with their quotation. They will usually provide unit price rates inclusive of overhead and profit for their works package. The contractor will use the rates with a suitable adjustment for each of the priced items.

The cost or output depends on many variables; attention should always be paid to the conditions that prevailed at the time when the particular recorded cost or output was noted, with consideration given to the levels of incentives that were used to achieve the particular standard. These conditions must be compared carefully with those expected to be encountered on the project under consideration. Differences between the estimated and actual cost or output on previous projects should be analysed, and

Table 4.2 Gang sizes with the relevant production and unit rates.

	Cost per hour (£)	Production hours	Unit rate for production (£)	Unit price rate per hour (£)
Ground works gang				
1 ganger	14.84	0.5	74.26	11.42
6 labourers	11.14	6.0		
Concrete placing gang				
1 foreman	14.84	0.5	65.52	11.92
5 general operatives skill level 3	11.62	5.0		
Bricklaying gang				
1 foreman	14.84	0.45	92.02	14.26
5 bricklayers	14.84	5.0		
1 general operative	11.14	1.0		
Painting and decorating				
2 painters and decorators	14.84	2.0	40.82	13.61
1 general operative	11.14	1.0		

any obvious conclusions noted. Adjustments must be made to update the estimating data.

Increasingly, there is more reliance upon the quotations provided by the specialty contractor who has developed knowledge and expertise in a highly specialised field. A concrete frame contractor will only focus on the formwork/shuttering and the fixing of the steel reinforcement and the concrete placing, hence will have good knowledge of the production standards they need to achieve to be competitive. This specialisation should lead to more reliable estimating. The contractor must make allowance for the management of the works package and the additional items required for the effective delivery of the works package.

When a particular type of work is being considered for the first time, there will be no previous cost or output records for guidance. Specialty/works contractors should be consulted whenever possible, and technical information from outside sources may have to be used. Information can be provided by manufacturers either through printed literature or technical representatives. Caution must be exercised when using any data from external sources.

Proper allowances must be made for any learning curve associated with new types of work and for incentive payments, either by increasing the all-in labour rate or by using an appropriately modified production standard. It may be necessary to adjust for any bonus included in the all-in labour rate.

Labour element

Labour costs are estimated on the basis of the all-in hourly rates previously established. It is recommended that 'gang costs' should be used for some trades in preference to individual hourly rates. Typically, an effective rate is built up for a member of a concreting gang or a bricklayer rate that will include a proportion of a labourer's time. However, it is also reasonable to price all trades on a normal hourly rate basis provided that an allowance for attendant ancillary labour is added to the established all-in hourly rates or added as an item of general labour in the project overheads stage.

It is usual to express outputs as 'decimal constants' such as 1.50 h to lay a square metre of brickwork. This is because computer systems conventionally expect estimators to rate items by inputting a resource code and a quantity.

Contractors and trade specialists assemble tables of data for use by their estimators as a guide to basic outputs. Many factors affect the time allowed for an operation or item, and careful consideration must be given to each of these factors, enabling the time allowed to be as accurate as possible.

The drawings, specification and bills of quantities

Items should be carefully examined to determine:

- Presence of any curved, splayed or unusual items that will slow the production rate
- Extent of standardisation of dimensions and layout
- Degree of accuracy and tolerances required
- Allowances needed for compaction, overbreak, batters and so on.
- Quality of finish and standard of workmanship required
- Whether operations are repetitive with the benefit of the learning curve in the production process

- Whether extensive and detailed specialist setting out will be required
- Whether the operation is likely to be within the experience of existing staff and operatives, whether special instruction or training is needed, or whether there will be a need to engage specially trained personnel
- The weight of specific items for materials handling and craneage
- Shift times, if shift work will be required to meet the programme schedule
- Requirement for contractor designed and installed work packages
- Any special local authority requirements in respect of noise, access and traffic.

The tender stage, method statement and logistics plan

The estimator must record any special factors, which lead to alteration of production standards to any considerable extent, in the Estimator's Summary and Report for consideration at the tender settlement stage.

This will show:

- Materials and components with long lead ordering times
- Load sizing for delivery of materials and components
- Restrictions on deliveries and unloading
- Any double handling of materials
- Special requirements for site security
- Special requirements for communication systems, including internet, intranet and cloud-based systems with collaboration tools to share information.

The tender stage construction programme and method statements

These will indicate:

- Time available for activities on the site, and activity duration for each of the activities, and their interrelationship
- Time of the year when work is to be undertaken and the likely seasonal conditions encountered
- What items can be pre-fabricated/assembled off-site and brought to the site for fixing
- Whether work packages will be continuous or intermittent
- Any restrictions that might affect normal working, such as restrictions on hours of working
- Degree of interdependence of trades, operations and work packages
- Restrictions in working, such as secure areas, safety and health requirements
- Environment in which the work will take place, such as hot/cold/exposed areas of work
- Facilities available for use by domestic and named specialty contractors as items of general attendance
- Pattern of production and the likelihood of achieving maximum possible rates

- Resources needed, such as the proportions of supervisory, skilled and unskilled operatives required, and recommended gang sizes
- Extent of mechanisation and off-site production envisaged and method of unloading, storing, handling and transporting materials.

The visit to the site and locality

The visit will have shown:

- Physical conditions and any likely restrictions
- Ground conditions with the site borehole reports, reliability of the borehole reports across the whole site, presence of rock and soil conditions
- Site layout, operating, storage and unloading facilities
- Special traffic requirements
- Site boundaries with any adjoining properties and any temporary protection for party walls
- Availability of skills, experience and availability of local labour. Presence of existing utilities and any restrictions on use
- The building control officer's requirements for notifications, approvals and visits
- Special temporary protection of any facilities
- Location of temporary offices for site huts and the need for any gantries. Location of the crane, plant and equipment
- Extent of scaffold required for the work.

4.7 Allocation of costs

When mechanical plant is used only on specific and limited operations (such as excavation and soil disposal), there is little difficulty in allocating the costs of the plant to specific items measured in the bills of quantities, taking into account the various factors noted above.

However, when an item of mechanical plant serves a number of trades or operations (e.g. a crane or hoist or a concrete mixer, which is used for concrete work and also brickwork and drainage work), then the allocation of its cost to measured items can only be made on an arbitrary basis. When the cost of an item of plant is associated with time on site rather than with specific items of measured work (e.g. pumping operations), then such items cannot reasonably be allocated against measured work.

In such circumstances, the cost of such plant, together with time in excess of productive output, must be included in the project overheads, rather than spread in an arbitrary manner over measured rates.

There are many examples of resources which are difficult to allocate with one cost category, such as falsework to support soffit. Formwork may be included in the material or plant element of a unit rate or equally can be assessed separately in project overheads as temporary works. For building work, the cost of falsework is commonly in the unit price rate; but this is not the case for civil engineering where all supporting equipment and structures tend to be linked to a resourced short-term programme and temporary works calculations.

Plant quotations

Plant hire enquiries must state:

- ✓ Title and location of the work, and address of the site
- ✓ Specification of the plant or work to be done
- ✓ Anticipated periods of hire with start date on site and duration required
- ✓ Means of access, highlighting any restraints or limitations
- ✓ Any traffic restrictions affecting delivery times
- ✓ Anticipated working hours of the site
- ✓ Date the quotation is required

The contractor must consider the intended method of working and any tender works programme requirements in the specification for plant. Turnaround of equipment and striking time will dictate the amount of formwork, support and access equipment needed. A balance must be drawn between speed of operation and economy in establishing plant needs, and all must be clearly reflected in the plant enquiry.

A list of plant suppliers must be established from companies who can meet the project's requirements. The options available for obtaining plant include:

- Purchasing plant for the project (in accordance with company policy).
- Hiring existing company-owned plant.
- Hiring plant from external sources.

Hiring existing company-owned plant

When plant is already owned by the company, the estimating department will be provided with hire rates at which plant will be charged to the site. The following list should be regarded as guidance only to the items which must be considered in building up hire rates for company-owned plant:

- Capital sum based on the purchase price and expected economic life (this will vary according to the company's accounting policy)
- Assessment of the costs of finance
- Return required on capital invested
- Grants and financial assistance available when purchasing plant
- Administration and depot costs
- Costs of insurances and road fund licences
- Maintenance time and costs and cost of stocks needed for maintenance purposes.

Purchasing plant for the project

The decision to purchase plant for a particular project is taken by senior management. Such a decision requires knowledge of plant engineering and will be made in accordance with the accounting policy of the company. Purchasing of plant is outside the scope of this code. However, for guidance purposes only, the manner in which the cost of such plant is charged subsequently to a contract will depend on the accounting policy of the company should be considered when plant is to be purchased for a project and sold on completion.

Hiring from external sources

Where company-owned plant is not available, enquiries must be sent to external suppliers for the plant required. Enquiries for items of plant must be sent either specifying particular machines and equipment that are needed or specifying the performance required from the item of plant.

In addition to the basic hire charge per hour or week, the enquiry must seek to establish:

- Period the quotation is to remain open.
- Whether fluctuating or firm price required, the basis for recovery of fluctuations and the base date when formulae are used for the recovery of fluctuations.
- Discounts offered.
- The person in the contractor's organisation to be contacted regarding queries.
- Cost of delivering and subsequent removal of plant from the site on completion of hire.
- Cost of any operator, over and above the basic hire charge, if provided by the hiring company. However, if provided by the contractor, the estimator must produce a built-up rate for the operator's costs.
- Whether the hire rates quoted include for servicing costs; if not, the costs and timing of servicing must be established.
- Any minimum hire periods applicable to the plant and the extent of any guaranteed time.
- Cost of standing time and insurance costs if the plant is retained on-site and not working for any reason.

4.8 All-in rates for plant and equipment

The plant requirements will be established in the method statement and programme. They will establish the basic performance requirements of the plant and in many cases will have identified specific plant items needed for the works. The duration for which the plant is needed on site will be established from the tender works programme.

The estimator must first compile a 'schedule of plant requirements', listing the type, performance requirements and durations. This should be separated into:

- Mechanical plant with operator
- Mechanical plant without operator
- Non-mechanical plant.

A note must be made on the schedule of additional requirements associated with a particular item of plant, which must be provided by the contractor. A power supply for a tower crane, for example could be a significant additional cost, and temporary access roads for erection purposes may be needed, together with foundations.

Further details are necessary for certain non-mechanical plant.

A scaffolding schedule must be drawn up by the estimating team in order to provide scaffolding contractors with a clear list of requirements. There are seldom work items

Table 4.3 Decisions when considering plant costs.

The manner in which time-related charges and fixed charges will be accommodated, that is delivery, erection and removal charges could be spread across the duration of hire and added to a weekly rate or, alternatively, shown separately as a fixed charge, separate from time- related costs in the project over heads	Rate of production likely to be achieved by the plant, bearing in mind the specific requirements of the project, the season of the year, and in the case of excavation work, the ground and water conditions
The continuity which can be expected for any item of plant and the likelihood of achieving a high production rate; it is unlikely that outputs quoted by manufacturers can be attained	Average output, making due allowance for intermittent working, site conditions, seasonal effects and maintenance

Source: CIOB (2011). CIOB's Code of Estimating Practice, 7th Edition (Wiley-Blackwell publication).

Table 4.4 Allowances to be made in establishing plant costs.

Divergence or discrepancy from the contractor's enquiry in the quotation which is being considered.	Delivery, erection and removal charges if applicable
Fuel costs, if applicable	Safety measures that are required
The effect and cost of maintenance and consequent down time of plant	Special provisions needed for unloading and loading plant
Temporary access roads, hard-standings or temporary works required for the plant	Weight restrictions which may affect the plant or its use
Whether any special insurances are needed for the plant, such as responsibility for the plant during delivery and erection	Availability of power for electrically operated plant; consider the need for a temporary sub-station or generators
Contractor's attendant labour requirements; banksmen are particularly important	Consents required for the use of plant on or over adjacent land
Supporting equipment needed to operate plant, e.g. crane slings, chains, skips, cages, etc., associated with lifting equipment, etc., associated with a compressor (these may be separately priced in the project overheads).	Allowance for damage, repairs and replacement parts chargeable to the contractor
Minimum hire charges	

Source: CIOB (2011). CIOB's Code of Estimating Practice, 7th Edition (Wiley-Blackwell publication).

in a BoQ for temporary works, although the preliminaries may give specific requirements such as temporary roof structures or bridging scaffolding to span low level obstructions.

The analysis of quotations received for plant will be set out in the plant comparison form and any additional factors to be priced identified. Allowances must be made for additional matters associated with the plant. In considering the total costs of plant, decisions must be made concerning the items shown in Table 4.3.

In all cases, the tender works programme requirements must reflect these conditions. In establishing the costs of plant, allowances must be made for additional matters associated with the plant – see Table 4.4.

The estimator must decide which items will be accommodated in the all-in rate for plant, where plant is to be allocated against unit rates, and which items are to be allocated in the project overheads.

Some examples of plant

CRANES

Mobile cranes

Self-propelled mobile crane on road wheels, rough terrain wheels or caterpillar tracks including lorry mounted:

- Max. lifting capacity at min. radius, up to and including 5T. Max. lifting capacity at min. radius, over 5T and up to and including 10T.
- Max. lifting capacity at min. radius, over 10T.

Tower cranes (including static or travelling: standard trolley or luffing jib)

- Up to and including 2T max. lifting capacity at min. radius. Over 2T up to and including 10T max. lifting capacity at min. radius.
- Over 10T up to and including 20T max. lifting capacity at min. radius.
- Over 20T max. lifting capacity at min. radius.

Miscellaneous cranes and hoists

- Overhead bridge crane or gantry crane up to and including 10T capacity.
- Overhead bridge crane or gantry crane over 10T up to and including 20T capacity.
- Power-driven hoist or jib crane with slinger/signaller appointed to attend crane or hoist to be responsible for fastening or slinging loads and generally to direct lifting operations.
- Pick and carry cranes for use in confined spaces without outriggers – with capacity up to 15T.

DOZERS

- Crawler dozer with standard operating weight up to and including 10T.
- Crawler dozer with standard operating weight over 10T and up to and including 50T.
- Crawler dozer with standard operating weight over 50T.

Dumpers and dump trucks

- Up to and including 10T rated payload.
- Over 10T and up to and including 20T rated payload.
- Over 20T and up to and including 50T rated payload.
- Over 50T and up to and including 100T rated payload.
- Over 100T rated payload.

Excavators (360° slewing)

- Excavators with standard operating weight up to and including 10T.
- Excavator with standard operating weight over 10T and up to and including 50T.
- Excavator with standard operating weight over 50T.
- Banksman appointed to attend excavator or responsible for positioning vehicles during loading or tipping.

Fork-lifts trucks and telehandlers

- Smooth or rough terrain fork lift trucks (including side loaders) and telehandlers up to and including 3T lift capacity.
- Over 3T lift capacity.

Power-driven tools

- Power-driven tools such as breakers, percussive drills, picks and spades, rammers and tamping machines.

Power rollers

- Roller, up to and including 4T operating weight.
- Roller, over 4T operating weight and upwards.
- Pumps, power-driven pump(s).

Shovel loaders (wheeled or tracked, including skid steer)

- Up to and including 2 m^3 shovel capacity.
- Over 2 m^3 and up to and including 5 m^3 shovel capacity.
- Over 5 m^3 shovel capacity.

Tractors (wheeled or tracked)

- Tractor, when used to tow trailer and/or with mounted compressor, up to and including 100 kW rated engine power.
- Tractor over 100 kW up to and including 250 kW rated engine power.

Trenchers (type wheel, chain or saw)

- Trenching machine, up to and including 50 kW gross engine power.
- Trenching machine, over 50 kW and up to and including 100 kW gross engine power.
- Trenching machine, over 100 kW gross engine power.

4.9 Select materials and specialty contractors' quotations

Materials quotations

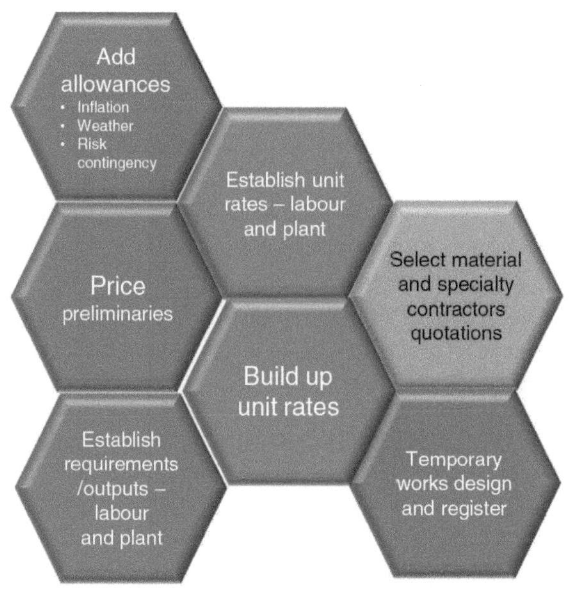

Enquiries to suppliers of materials should state:

- Title and location of the work and site address
- Specification, class and quality of the material
- Quantity of the material and the load sizes for delivery
- Likely delivery programme, that is period during which supplies would be needed with daily or weekly requirements where known; where small quantities are to be called off from a bulk order, this must be stated clearly
- Means of access, highlighting any limitations or delivery restraints, and any traffic restrictions affecting delivery times
- Special delivery requirements such as palleting or self-unloading transport
- Date by which the quotation is required

- Period for which the quotation is to remain open
- Whether fluctuating or firm price required, the basis for recovery of increased costs and the base date when a formula is used for calculation of fluctuations
- Discounts required.

The person in the contractor's organisation is responsible for queries. The contractor has a responsibility to ensure that suppliers:

- Make every effort to meet the specified 'date required'. If this is not possible, the contractor must ensure that he is informed promptly so that additional enquiries can be sent out in order to maintain a full enquiry list
- Clear queries as they arise in order to avoid quotations marked 'more information required'
- Submit the quotation on time with a clear statement where prices are 'to follow'.

Particular attention should be paid to material supply conditions which may have cost implications. The contractor can incur costs for:

- Pallets left on site and any deposits not refundable
- Standing time for vehicles while unloading at the site
- Small quantities or abnormal loads.

Consideration should also be given, and the cost should be established, for various additional matters associated with materials. These include:

- Any specific divergence or discrepancies from the contractor's enquiry in the quotations received from the supplier
- Any minimum delivery requirements and adjustment of cost due to delivery in small quantities
- Trade discounts, which should be noted separately and reported at the settlement meeting. (Note that discounts may or may not be deducted from the materials cost at this stage. Some contractors maintain that materials costs should be net of discounts, which are summarised in the estimate summaries. Others allow the discount to remain in the materials cost but recognise the element when considering the profit mark-up at the settlement stage.)

Waste allowances

The allowance made for waste must, wherever possible, be based on experience gained on previous projects. Data given in textbooks, periodicals and manufacturers' catalogues should be examined critically and used with caution. The waste allowance must be carefully applied according to the circumstances of the project and previous experience of the material. Particular attention should also be given to any implications of the environmental regulations, see ***Environmental Protection*** honeycomb in the Principles section. Waste can also be an issue where a material has a minimum load size, leading to surplus material being re-stocked with a re-stocking charge, or taken into store in the hope it can be used on future projects.

Unloading, storage and distribution costs

The degree of mechanisation in unloading must be considered, to ensure that material deliveries are compatible with the intended method of handling, such as palleted materials for handling by forklift truck. Special equipment should be considered for unloading, although the costs of skips, slings and chains are more likely to be costed in project overheads than allocated directly against unit rates. The labour costs of unloading and distributing materials must be considered and an allowance made when establishing the total labour requirements of the project. Such labour can either be taken into account when selecting production standards for labour or be priced as a project overhead. Items to be accommodated include:

- Storage needs and protection
- Size and weight of materials
- The cost of any special packaging and crates, if these are charged, or the cost of returning them to the supplier
- Any subsidiary fixing materials or temporary materials needed for storage
- The anticipated time and rate of delivery required for materials and the amount to be stored on site, the location and method of subsequent distribution – both of these will be indicated in the tender works programme and method statement
- Unloading point.

The identification and costing of these various factors and considerations will convert the basic cost contained in the quotation into the cost which will be inserted into the net unit rates.

4.10 Specialty contractor quotations

Most contracts have provision for the contractor to sub-let work. The analysis of quotations received for domestic specialty contractors identifies any further matters which have to be costed by the contractor. Selection of the specialty contractor to be used may not be possible before such additional costs have been determined.

Allowances must now be made for any additional matters associated with the domestic specialty contractor's works, including:

- Specific divergence or discrepancy from the contractor's enquiry included in the quotation.
- Allowance for unloading, storage and protection of materials and equipment and transfer of goods from stores to point of work, if this is to be the main contractor's responsibility. The labour costs associated with unloading and distribution of materials are considered at this time and allowance made when establishing the total labour requirements of the project. Such labour is either taken into account by an addition to the sub-contractor's quotation or can be priced in the project overheads.
- General attendance items to be provided by the main contractor.

In making such allowances, the contractor must take into account the requirements of the tender works programme and method statement and facilities which have already been allocated for the contractor's own works.

Additions to cover attendance of domestic specialty contractors may be done in several ways, by:

- Increasing the relevant unit rates of the specialty contracted work;
- Adding a fixed percentage to the whole of the specialty contractor's quotation and
- Making an addition subsequently in the project overheads.

Discounts

Discounts offered by specialty contractors must be noted separately and reported at the settlement meeting. Discounts may or may not be deducted from the specialty contractors' quotation at this stage. Some contractors maintain that specialty contractors' costs should be net of discounts, which are summarised in the summary reports. Others allow the discounts to remain in the cost of the work to be sub-contracted but recognise the element when considering the profit mark-up at settlement stage.

Great care must be taken in assessing specialty contract or quotations to ensure that all items have been adequately covered. If labour-only specialty contractors are being considered, the cost allowance must take into account all factors associated with the provision of materials by the contractor, and adequate safeguards must be made to control the use and wastage of such materials.

4.11 Provisional sums – defined and undefined

The defined provisional sum will be included in the bid price, with due allowance for the influence of the work on the sequence and timing of operations. The bill of quantities/ specification should include statements about how and where the (defined) work is to be undertaken and an indication of the scope and extent of that work.

Provisional sums are included in bills of quantities for items of work which cannot be fully described or measured in accordance with the rules of the method of measurement at the time of tender. For work measured under the rules of NRM2, there are three types of provisional sum as listed below.

Provisional sum for defined works

This provisional sum is used where works are known to be required in the project but have not been fully designed or specified at tender stage and so cannot be measured in detail. The contractor must make due allowance for the planning engineering, project scheduling and pricing preliminaries; to enable him to do so, the following information must be provided with the provisional sum:

- the nature of the work
- how and where it is to be fixed
- quantities showing the scope and extent of the work
- limitations on method, sequence and timing.

Provisional sum for undefined works

An undefined provisional sum is for work that has not been designed and the scope and extent of that work is unclear. The undefined provisional sum amount will be included in the bid price.

Where the information required in support of a defined provisional sum is not available, the provisional sum is 'undefined', and the contractor is not required to include any duration for the work in the tender works programme, nor to take account of the costs of planning engineering, project scheduling or preliminaries. Undefined provisional sums are typically used to make contingent provision for possible expenditure on elements of work which cannot be wholly foreseen at tender stage or cannot be quantified. A client's contingency sum is deemed to be an undefined provisional sum.

Provisional sum for works by statutory authorities

NRM2 makes provision for a provisional sum to be included in a BoQs which is neither 'defined' nor 'undefined', for work to be carried out by the local authority or statutory undertakings, including privatised services authorities carrying out statutory works.

4.12 Incorporating provisional sums in an estimate

Although not strictly required by NRM2, it is common for items to be measured to enable the contractor to price attendance and profit on works by utilities providers as though they were nominated specialty contractors. It should be noted that these provisional sums are net and do not include a contractor's discount.

Adequacy of information

The rules of measurement give the items to be included in the bills of quantities for each named specialist contractor, as follows:

- The nature and construction of the work
- A statement of how and where the work is to be fixed
- Quantities which indicate the scope of the work
- Any employers' limitations affecting the method or timing sequence or sectional completion of the works
- General attendance items
- An item for main contractor's profit, to be shown as a percentage
- Details of special attendance required by the sub-contractor.

The estimator must check that the measured items for works which are covered by a PC sum are adequate and that supporting details are available in accordance with the appropriate method of measurement.

Named suppliers

Prime cost sums define the rate for material supply from suppliers, such as price per square metre for floor tiles.

Methods of measurement state that the cost of materials from named suppliers is identified in the tender documents as PC sums. A separate item is also given for the contractor to add profit. PC sums may also be written into an item description (such as a rate for the supply of facing bricks) for the estimator to incorporate the cost in the rate build-up. Named suppliers effectively become a domestic supplier and would not be identified separately in the tender documentation for costing purposes.

The estimator will produce a list of named suppliers and specialty contractors at an early stage using the schedule of PC sums and attendances. Where details associated with a nominated supplier are unclear, the estimator must note any concerns in their report for further consideration at the settlement meeting. If a PC sum has been included for high-value materials or large quantities, the estimator must check:

The terms of the purchase contract to check if any discounts have been allowed by the suppliers. This discount is normally deducted from the quotation in order to include net costs in the summaries for the settlement meeting
Delivery times and how they affect the works programme
Fixing items associated with materials provided by a named supplier need to be described adequately and measured in the items to be priced. Any discrepancies concerning fixings, such as bolts, screws, brackets, adhesives and sealants, or ambiguity over the responsibility for supply of these items must be clarified
Additional costs for unpacking, storage, handling, hoisting and the return of reusable crates or pallets to the supplier. Suppliers may deliver their materials in reusable crates or other packaging and the contractor may be required to return such items to the
Due allowance must be made for the collection, storage, handling and subsequent dispatch of such items back to the supplier
Where bills of quantities are used, the fixing of materials supplied by named suppliers is measured in the appropriate part of the bill

Named specialty contractors – named specialists

A named specialty contractor arises where the selection of a specialty contractor is to be made by the client or his representative, for which a sum has been inserted in the tender documents. The contract should be checked to ensure that this is permissible and that supplements have been provided if no main contract clauses exist.

There is a right of reasonable objection to a particular named specialty contractor because it would be contrary to contract law for a party to be forced into a contract against their will, where they had reasonable cause for not wishing to contract with a particular nominated party. Where PC sums are included in bills of quantities, the estimator is given the name of the named specialty contractor to discuss methods and project planning issues prior to tender.

The use of named specialty contractors is becoming increasingly rare for two reasons:

- The growing complexity of contractual procedures
- The risks carried by a client when a contractor is relieved from a proportion of the responsibility for full performance.

Where naming is used, they can sometimes account for a significant proportion of the overall cost of a contract. Contractors are frequently given inadequate supporting information to deal with attendances in the bill of quantities.

The adequacy of the information provided must be carefully investigated, and further particulars requested by the estimator if details are not complete. This enables the estimator to be able to include dates, nominal design, fabrication, delivery work period dates and interfaces for work in the tender works programme.

Attendances

Attendance is defined as, 'the labour, plant, materials or other facilities provided by the main contractor for the benefit of the specialty contractor and for which the specialty contractor normally bears 'no cost'. The main contractor is responsible under the main contract provisions for the site establishment and providing attendance. This provides clear responsibilities for the support services and equipment needed on site and eliminates duplication of resources for various specialist sub-contractors.

For very large contracts, where a construction manager or management contractor has overall control, trade/works contractors are asked to provide certain parts of the temporary works and facilities themselves. The costs associated with attendance are built into the main contractor's tender and consequently become a charge against the client. However, the associated risks of attendance are borne by the main contractor.

The estimator must decide how to price 'general attendance' and 'special attendance' relating to nominated sub-contractors. The attendances may be priced in the project overheads schedules or on the schedule of PC sums and attendances.

General attendance

The item for general attendance is an indication of the facilities which are normally available to specialty contractors and that they are provided by the contractor to meet regulatory and logistical requirements. In assessing any sums to be allowed for general attendance, the estimator must investigate the facilities which will already be provided for the main contractor's use and determine any costs which may arise by the named specialty contractors' use of any such facilities. Table 4.5 shows the facilities that may be required.

Table 4.5 General attendance facilities.

Use of temporary roads, paving and slabs. Allowance must be made for any costs associated with the maintenance of temporary roads, paving and paths that are required during the time period allowed by the contractor for his own use. This item will not cover any specific access requirements of a nominated sub-contractor. For example, such items as hard standing for a crane should be separately described under 'special attendance'	**Use of standing scaffolding**. The contractor must allow for any costs which might arise through the named specialty contractors' use of scaffolding which is already erected for the main contractor's use. Any modifications or additional scaffolding required, or, any increase in duration for such scaffolding over and above the time period required by the main contractor, must be described and measured as 'special attendance' in the bills of quantities
Use of standing power-operated hoisting plant. While named specialty contractors may use existing hoisting plant if there is spare capacity, any hoisting facilities specifically required must be measured under 'special attendance'	**Use of mess rooms, toilets and welfare facilities** Assessment must be made of the accommodation needed for the operations of named specialty contractors over and above the requirements of the contractor. Allowance must also be made for any servicing and cleaning of such facilities which are shared with the contractor.
Provision of temporary lighting and water supplies. The estimator must establish requirements for general lighting needed to comply with safety requirements and for the execution of the works during normal working hours. Adequate allowance must be made for water points needed for the construction of the works. This may mean the simultaneous provision of such services in other areas of the building over and above the requirements of the main contractor. Special lighting requirements and power needs must be measured and priced under 'special attendance'. Specific water requirements for testing or associated with commissioning of plant should be measured under 'special attendance'	**Special scaffolding**. In order to price this item, the estimator must be given precise details concerning the scaffolding requirements. Such information should define clearly the height in stages of the scaffolding, indicate the extent of boarded platforms and any alteration and adaptation that will be required. If such information is not available and descriptions are inadequate, the estimator should seek further instructions from the consultant. The estimator must also make due allowance under this heading for any adaption or alteration to standing scaffolding or for any extension to the time period, providing such items are described and measured in the bills of quantities
Unloading, distributing, hoisting and placing in position. This may include some intermediate storage requirement and due allowance should be made for this. It is essential that particulars are stated (e.g. size or weight) of materials to be handled to enable the estimator to reasonably assess costs and identify the appropriate mechanical aids. With heavy units, such as precast cladding, it will be necessary to know the delivery rate and any specific stacking facilities required for site storage. Sufficient information must be provided to identify any distribution requirements, as opposed to hoisting and stacking. In assessing the cost, the estimator must consider the use of existing mechanical hoists and ensure that sufficient hoisting capacity is available. The estimator may need to seek clarification if components (e.g. precast concrete cladding panels) are to be placed in position, as there could be overlapping responsibilities with the specialty contractor who would be expected to supply and fix all of their own materials	**Special attendance.** Other specific attendances which do not fall under the category of 'general attendance' must be specifically measured in the bill of quantities as 'special attendance', these include: ■ Special scaffolding or scaffolding additional to the contractor's standing scaffolding ■ The provision of temporary access roads and hardstandings in connection with structural steelwork, precast concrete components, piling, heavy items of plant ■ Unloading, distributing, hoisting and placing in position, giving, in the case of significant items, the weight, location and size ■ The provision of covered storage and accommodation, including lighting and power thereto ■ Power supplies giving the maximum load ■ Any other attendance not included in 'general attendance' or listed above
Provision of covered storage and accommodation including lighting and power. Under 'general attendance' the contractor is required to provide space for nominated sub-contractors to erect their own facilities. Under this item the main contractor will be required to provide, erect and maintain accommodation and provide lighting and power as stipulated. The size of hutting required should be stipulated and the period required stated. Any special requirements, that is racking or other services, should also be defined	**Clearing away rubbish.** The disposal of waste, packaging and other rubbish from an agreed collection point involving labour, containers and haulage must be assessed. Abnormal items of rubbish, such as disposing of surplus excavated material from a ground improvement technique, must be measured separately under 'special attendance'. The prevailing regulatory requirements should also be taken into consideration (e.g. regarding disposal of hazardous waste material)

(Continued)

Table 4.5 (Continued)

The provision of temporary access roads and hardstandings. Where any specific requirements are described, these must be taken into account with the contractor's own needs, and any additional temporary provision allowed for

Power supplies giving the maximum load. Any special power requirements, including power for testing of systems, must be clearly measured for pricing purposes. Any reference to power supplies should state whether single or three-phase electrical power supplies are required, and the maximum demand level should be taken into account. The estimator should ensure that any descriptions for fuel or power for such testing purposes are clearly specified, giving the quantity necessary to fulfil the tests and also the precise specification of the power needs

Maintenance of specific temperature or humidity levels. Any specific requirements for controlling temperature or humidity must be clearly measured, stating temperature/humidity required and the time period for the provision of these services. The requirement must also state if the permanent services in the building can be used for this purpose

Providing for specialty-contractor's own space. The estimator should note that only space is required and that cover in the form of a shed is not a requirement. The assessment of total space requirements must be borne in mind when finalising the method statement and site layout

Any other attendances. The contractor is required to provide specific attendance or materials for various trades. This could include the provision of bedding material for roof tiles, or floor tiles. Other items such as specific cleaning operations and the removal of masking tape used by subcontractors should also be defined. Such items must be clearly measured, in the event of any inadequacies or ambiguities, the estimator should refer to the consultants for further instructions

Source: CIOB (2011). CIOB's Code of Estimating Practice, 7th Edition (Wiley-Blackwell publication).

With the exception of provisional sums for work by statutory authorities, provisional sums are deemed to include an allowance for the main contractor's head office overheads and profit. All provisional sums will become the subject of an architect's instruction during construction, and the work will be valued according to the appropriate contract rules for measurement and valuation which include provision for overheads and profit.

Accordingly, the value of provisional sums (except those for statutory authorities) should be added into the final summary after the application of overheads and profit. Alternatively, if it is company practice to add head office overheads and profit to the total value of measured and unmeasured work. Provisional sums should be discounted before entry in the summary to avoid duplicating the overheads and profit for this type of work.

4.13 Daywork

Definition

Contractors must understand the circumstances in which varied or additional work will be valued on a daywork basis. It normally occurs where variations cannot be valued by measurement using unit price bill rates or comparable rates, nor by negotiation before an instruction is issued.

The daywork charges are usually calculated using the definitions for PCs and overheads published by the RICS/BEC for building work and FCEC for civil engineering. The PC of daywork can be defined in other ways, so care must be exercised in reading the definition in the tender documents. The composition of the total daywork charge will include the following costs:

- Labour
- Materials and goods

- Plant
- Supplementary charges (civil engineering contracts)
- Incidental costs, overheads and profit (this addition will vary between labour, plant and materials and, in order to introduce competition at tender stage, is added to provisional sums for the PC of labour in the bills of quantities by the contractor).

An alternative method (for labour to be valued on a daywork basis) is for the contractor to provide all-in gross hourly rates which are applied to provisional hours. This makes the calculation of daywork rates simpler during the course of the project but moves the burden for anticipating increased costs to the contractor.

Contractors may decide that some of the project and head office overheads are covered in the contract price and may be excluded from daywork rates. This is mainly true if the daywork, to be carried out during the currency of the contract, will not result in an extension to the contract, but other additional costs to project and head office overheads may still have to be considered.

It is inappropriate to use this payment method for anything except work which is incidental to contract work. In the event that significant changes are made to the original scope of works, the valuation rules normally allow additional overhead costs to be recovered, usually when the full effects of changes are known.

Decisions concerning allowances for profit and overheads must be made by each contractor taking into account their own circumstances and method of working and his assessment of the effects of daywork on a particular project. The contractor must assess each contract on its own merits in producing daywork rates and calculating the percentage addition needed. This will include an assessment of the likelihood of the PC being a reasonable pre-estimate of the work, which will be valued on a daywork basis.

The contractor's daywork percentages must take into account the rates required by the specialty contractors used in the tender. Enquiries to specialty contractors must include a request for daywork percentages based on the definition incorporated in the main contract. For mechanical and electrical installations in building contracts, the contractor is given the facility to state different percentages for specialists in the bills of quantities. As dayworks are calculated inclusive of an allowance for overheads and profit, they should, like provisional sums, be added into the final summary after the application of overheads and profit.

4.14 Pricing the preliminaries

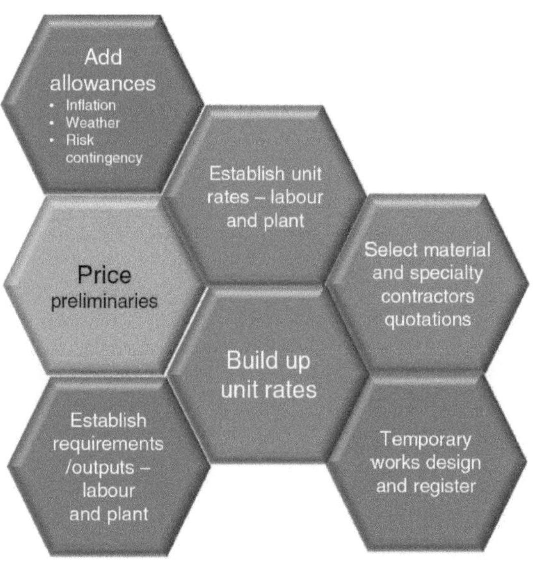

The preliminaries are a significant cost item and should be priced carefully. However, as they occur at the beginning of the project, cost estimates are approximations and should be checked often as more information becomes available. More details about preliminaries are in Chapter 8 where the Preliminaries honeycomb itemises the different elements involved.

4.15 Add allowances

Inflation

Add allowances
• Inflation
• Weather
• Risk contingency

Establish unit rates – labour and plant

Select material and specialty contractors quotations

Price preliminaries

Build up unit rates

Establish requirements /outputs – labour and plant

Temporary works design and register

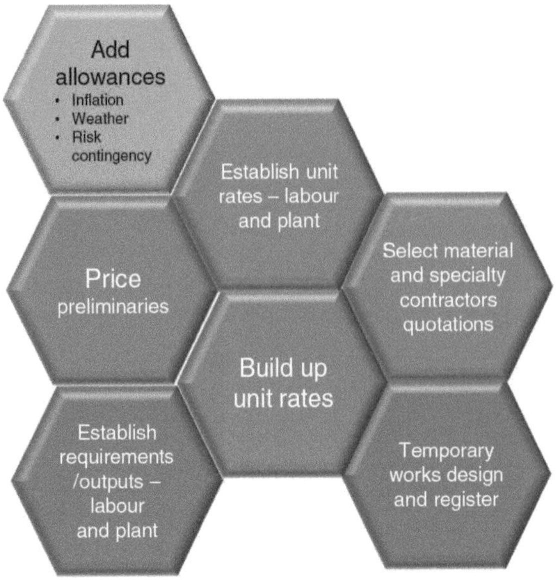

In times of low and predictable inflation, contractors are expected to submit tender prices that remain fixed for the anticipated duration of the work. Fluctuations will be paid under a fixed priced contract in certain circumstances, for example when materials are purchased during a period of compensable prolongation of the contract. For these purposes, the basic cost of materials is recorded in a list in an appendix to the BoQ.

Where the contract is to be fully fluctuating, the method of establishing the increases or decreases in cost need to be checked. It is common for a formula method to be adopted, and the rules for the application of the formula need to be studied carefully.

Where a firm price is to be submitted, an assessment must be made of the likely variations in cost during the proposed contract period. The tender works programme is an important tool in assessing the likely impact of price rises on certain elements of the project. There are some parts which will not need consideration such as provisional and PC sums and firm price quotations which fully comply with the conditions of contract.

Risk allowance

Risk allowance – *the amount added to the base cost estimate for items that cannot be precisely predicted to arrive at an allowance that reflects the potential risk.*

This allowance may be calculated as a percentage of the capital cost of a project. In public projects, a quantitative risk analysis (QRA) is used to generate a risk allowance. This method focuses on the confidence level of the risk allowance not being exceeded. An 80% confidence level is mostly used and is referred to as the P80 risk allowance. There is software available to undertake QRA, using techniques such as Monte Carlo simulation. The QRA process informs the risk allowance process rather than producing a risk allowance figure.

The number of days of weather-caused delay in money terms allowed by contractors in their returned tenders is less than the actual number of days of inclement weather affecting the progress of works. This indicates that contractors generally absorb some of the delays in their program or absorb some of the delaying costs in their intended profits Chan and Au (2008, p. 679).

Risk allowances can be split into three levels: (1) compensation for inaccuracies/errors in the estimating process; (2) cover for identified risks and (3) the residual risk allowance which is decided by management based on the need to win the project. A response to risk in the tender needs to be carefully considered by the bid team in consultation with others. The likelihood and impact of each risk needs to be quantified. There are three possible responses:

- Risk transfer to the contractor;
- Risk sharing by both employer and contractor;
- Risk retention by the employer.

NB Risk is covered in more detail in the Principles section.

Weather and the estimating process

Weather and climate have a huge impact on a construction project, affecting time, cost, productivity, health and safety, quality and plant performance. Making allowances for the impact of weather is particularly important at the estimating stage when pricing. The bid will be based on the tender documents yet weather/climate risk is not usually one of the BoQ items (Chan and Au, 2008); there is an expectation that the contractor will add allowances for the impact of inclement weather.

Productivity, performance and health and safety (Moselhi *et al.*, 1997) can be compromised by adverse weather, as can materials, plant and equipment. Working at heights and the use of tower cranes are examples of where time, cost and safety would be compromised by high winds. Table 4.6 shows the impact on different activities in a range of weather conditions.

High winds

Sites that are exposed will suffer more from high winds both in the danger to labour on-site and crane and hoisting operations. The contractor's dust control measures would also be affected. The estimator would need to allow for extra bracing of partially completed structures and, in the case of severe winds (gale/hurricane force), structures may need to be tied down, sandbagged or materials moved off-site into a covered facility.

Concrete

Concrete can be poured in temperatures as low as 4 °C, but low temperatures create extra costs which need to be taken into account by the estimator, such as:

- Cost of admixtures
- Formwork insulation
- Removing ice from formwork
- Protecting newly poured concrete.

Conversely, when the weather is hot (over 26 °C), steps need to be taken to ensure that the concrete is maintained at a lower temperature by using cooled/iced water/liquid nitrogen for the mixing process. The use of admixtures and low-heat cement can help. All these precautions and procedures increase the cost of pouring, placing and curing concrete (Sinclair *et al.*, 2002).

Table 4.6 The impact on different activities in a range of weather conditions.

Activity	Heavy precipitation	Drought	Strong wind	Lightning	Low temperature	High temperature
Ground clearance	✓		✓	✓		
Concrete	✓		✓	✓	✓	✓
Labour schedule	✓		✓	✓		✓
Drainage	✓	✓		✓		
Embankment construction	✓	✓	✓	✓	✓	
Erosion and sedimentation	✓		✓	✓		
Excavation	✓			✓	✓	
Fencing	✓		✓	✓	✓	
Painting	✓		✓	✓	✓	✓
Paving	✓				✓	
Steelwork	✓		✓	✓		
Soft landscaping	✓	✓	✓		✓	✓

Source: Transportation Research Board (2014). CIOB's Code of Estimating Practice, 7th Edition (Wiley-Blackwell publication).

Other materials

- *Mortar*: Dry weather can cause mortar to dry prematurely, reducing body strength.
- *Paint*: The weather (temperature and humidity) can affect both the performance and application of paint.
- *Seals and sealants*: Freeze–thaw cycles and UV exposure will cause a loss of elasticity, with seals and sealants becoming brittle.
- *Wood, insulating materials and plasterboard* are adversely affected by wet weather.
- The provision of temporary protection is one solution and the cost of doing this – materials, labour equipment for the erection and dismantling – will need to be allowed for in the bid.

Weather risk management

Other Met Offices
USA - National Weather Service World Meteorological Organization with 185 Member States and 6 Territories

Meteorological organisations provide reports and assessments that can inform across the planning, production and operation phases. For example, the UK Met Office provides location-based reports for 3,600 locations that include monthly planning averages and monthly downtime summaries. The planning averages are based on a 30-year period to provide 'seasonal norms'. The downtime summaries provide the detailed weather conditions experienced within a particular month(s) of interest and compare this to the long-term averages (LTAs) and 1-in-10year values. This is particularly important in the case of contractual disputes. Figure 4.3 show the services provided by the Met Office over a project's lifetime.

Many of the more detailed services have to be bought. Figures 4.4 and 4.5 give examples of the reports provided.

Contracts and weather

The estimator must review the contract clauses related to weather delay. If there is none, then the client's position on such events should be sought regarding notice, extensions and compensation.

Clients will usually allocate risk of delays due to adverse weather contract provisions such as 'weather', 'default' and 'force majeure' clauses. An extension of time to a contractor for a delay caused by abnormally adverse weather conditions is typically given.

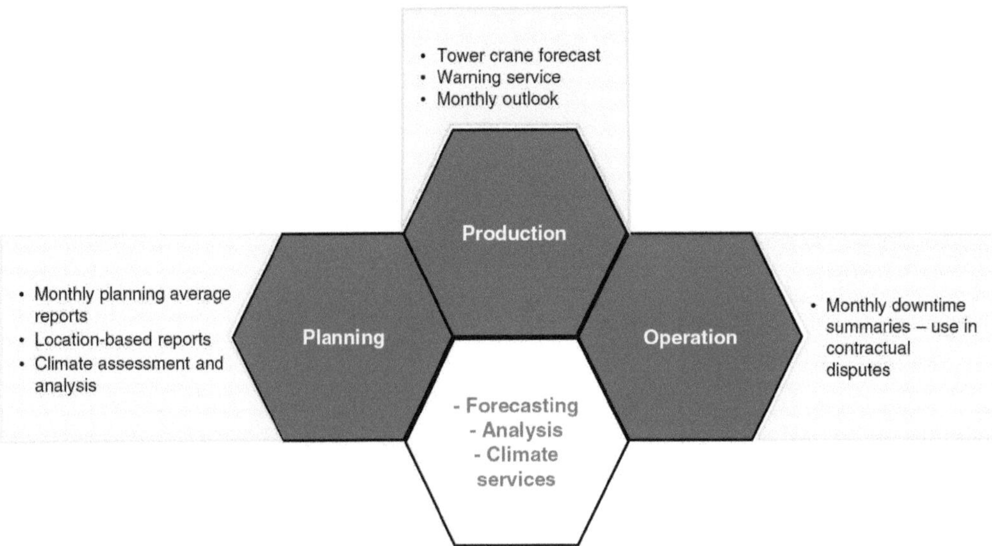

Figure 4.3 The services offered by the UK Met Office.

Under a Joint Contracts Tribunal (JCT), contract contractors are entitled to an extension of time for event caused by 'exceptionally adverse weather conditions', but does not define the term. Thus, it is open to the discretion of the contract administrator and so can be the source of a dispute.

NEC contracts have a definition – 'where the weather over a calendar month has occurred on average less frequently than once in ten years'. Under NEC3 the contractor is entitled to a change in the prices and the reasonably foreseeable costs of dealing with the adverse weather. International Federation of Consulting Engineers (FIDIC) contracts make provision for 'exceptionally adverse weather conditions'. Where 'exceptionally adverse weather conditions' occur, the risk is shared: the contractor will benefit from an extension of time but recover no loss and expense; the client recovers no liquidated damages.

Location based downtime summaries

Prepared for:	Exeter
Site:	Postcode EX1 3PB
Weather data from:	Latitude 50.7242, Longitude –3.5047
Month:	January 2015

Issued on Thursday 5 March at 12:34:45

Summary Page

	Monthly summary for January 2015	1-in-10 year value (1971–2010)	Long-term average (1981–2010)
Monthly rainfall total (mm) 0900-0900	85.1	149.0	85.0
Total days of rain > 5 mm	5	11	6
Monthly snowfall total (cm)	0.2	Not available	Not available
Total days of snow	2	5	2
Total days with snow lying at 0900	3	4	1
Maximum snow depth (cm) at 0900	0.0	Not available	Not available
Total days of freezing	0	2	0
Minimum monthly ground temp (°C)	–3.2	Not available	Not available
Total days of ground frost	4	22	15
Minimum monthly air temp (°C)	–4.7	–7.8	–4.0
Total days of air frost	7	15	8
Mean monthly wind speed (mph) 0900-0900	10.7	13.0	9.8
Monthly sunshine total (hours)	57.2	78.0	58.0
Maximum monthly gust speed (mph) 0900-0900	63.3	Not available	Not available
Total days of lightning	0	Not available	Not available
Monthly solar radiation total (kWh/m^2)	31.9	30.0	26.0

Monthly value is less than or equal to the 1-in-10 year value, except for minimum ground temp and minimum air temp where the monthly value is greater than or equal to the 1-in-10 year value.

Monthly value is greater than the 1-in-10 year value, except for minimum ground temp and minimum air temp where the monthly value is less than the 1-in-10 year value.

Figure 4.4 An example of a location-based report.

Station Based Downtime Summary for ABERPORTH

Weather data from: (Lat = 52:14N Long = 06:22W)

Month: January 2014

Issued on Monday 1 December 2014 at 11:42:40

Date	Daily rainfall total (mm) 0900–0900	Days of rain >5mm	Minimum air temp (°C)	Days with air frost	Snow depth (cm) at 0900 UTC	Days with snow lying at 0900 UTC
01	3.8		3.3		0	–
02	3.2		3.3		0	–
03	0.6		3.5		0	–
04	1.8		2.6		0	–
05	7.8	1	0.4		0	–
06	0.2		4.7		0	–
07	0.6		6.6		0	–
08	0.2		4.3		0	–
09	0.2		–0.1		0	–
10	4.4		2.7		0	–
11	0.0		0.3		0	–
12	3.8		0.4		0	–
13	0.6		1.5		0	–
14	9.6	1	−1.4		0	–
15	2.8		–0.2		0	–
16	17.6	1	4.1		0	–
17	tr		4.3		0	–
18	9.4	1	3.3		0	–
19	0.2		2.4		0	–
20	2.4		1.7		0	–
21	4.4		1.8		0	–
22	2.0		4.8		0	–
23	5.0		2.9		0	–
24	6.4	1	2.5		0	–
25	16.2	1	4.9		0	–
26	2.0		2.1		0	–
27	7.8	1	2.5		0	–
28	0.8		4.6		0	–
29	2.6		3.8		0	–
30	2.0		2.9		0	–
31	17.4	1	3.3		0	–
Total	136.4	8	–	3	–	–
1–in–10 year value	126.7	10	–	16	–	7
Long–term average (1981–2010)	81.3	6	–	9	–	2

Figure 4.5 An example of a station-based downtime summary.

5 Prepare estimator's report

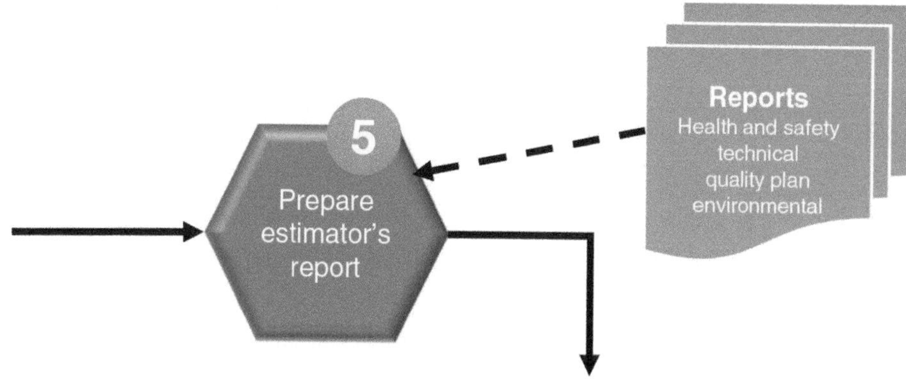

New Code of Estimating Practice, First Edition. The Chartered Institute of Building.

It is essential that all pertinent facts which have an influence on the settlement are presented in a logical and structured sequence.

During the estimate preparation, a large number of individual constituents will have been considered, including:

- *The project and its construction characteristics*: identifying the market sectors, the size, nature and construction of the building and any special features
- *Contract conditions*: identifying the authorising body,for example Joint Contracts Tribunal (JCT), a statement on design responsibility and a period for which the contract is to exist
- *Method statement*: describing the sequence, interface dates for commencement, completion and any phased handover or sectional completion and any resource or preferential logic
- *Tender works programme*: setting out the activities, durations and logic of the bar chart/critical path network. Identifying the sequence, interface dates for start, finish and any phased handover or sectional end dates
- *Specialty contractors and suppliers*: including details of any specified suppliers and products that may be from a single source
- *Insurances*: identifying the particular clauses of the contract which will apply to the works, together with a statement of requirements in terms if insurances are required to be held by the contractor
- *Site factors*: identifying access and egress, ground conditions, environmental considerations and any constraints or controls to normal operational processes
- *Bonds and warranties*: identifying the level and period for bonding and the number and type of warranties to be given and by whom (usually including specialty contractors with a design input)
- *Contractor's own preliminaries costs*: identifying plant, equipment and staffing numbers necessary to properly carry out and supervise the works
- *Health, safety and environmental issues*: identifying those that may be relevant to the particular site, to particular components within the construction or constraints imposed by adjoining properties
- *Cash flow*: will indicate whether the project will be cash positive or whether financing will be a consideration
- *Client issues*: which may identify their position in this particular marketplace together with trading history and payment performance
- *Consultant issues*: looking at the size, capability and design experience of the particular practice in that market sector together with the numbers and experience of personnel the practice can dedicate to the project
- *Pricing*: looking at the competition, the period for tendering and the situation in the marketplace in terms of available work and available resources to build
- *Procurement matters*: relating to lead-in times for specific trades and/or goods where the design requires approval by the consultants before the purchase and fabrication of raw materials
- *Quality*: this may be benchmarked in recognisable terms of specification or by example or sample of products and finishes

- *Design*: looking at the complexity of the scheme, the design of interfaces and the available pool of skills currently available in the sub-contract market
- *Risk/opportunity*: this is in effect a summary of the factors listed above reflected in the settlement either in financial terms or by an initial qualification.

Some of these items will have been factored into the decision to tender, and it is important to review these in that context in order to confirm that the final product still fits with the requirements of the company. For example, an organisation might have secured projects during the tender period that have an impact on the available resources making the current opportunity less attractive, that is the company may have to recruit staff for the project which would add a degree of uncertainty in outcome. All these items will be considered by management in relation to four key aspects of settlement.

Where there is a specific need for contractor design to be included on a traditional project, an allowance for this risk in relation to the legal liabilities of the design needs to be included. A designed portion supplement would be used in the contract to transfer design liability, and this would give the value and scope of the work to be covered. The risks associated with the designed portion would need to be brought to the attention of the management for consideration at tender settlement stage.

Although the final decisions are made by management, those concerned with estimating, planning, construction and commercial management and purchasing must be encouraged to communicate the knowledge they have acquired throughout the estimating stage to the review panel. Their contribution may be by attending the meeting or reporting through the estimator.

Examine and consider

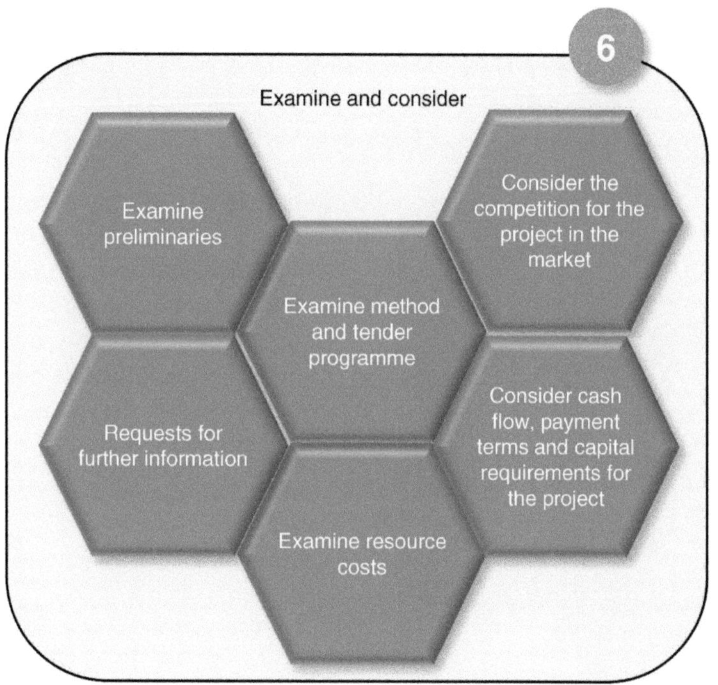

New Code of Estimating Practice, First Edition. The Chartered Institute of Building.

6.1 Examine preliminaries

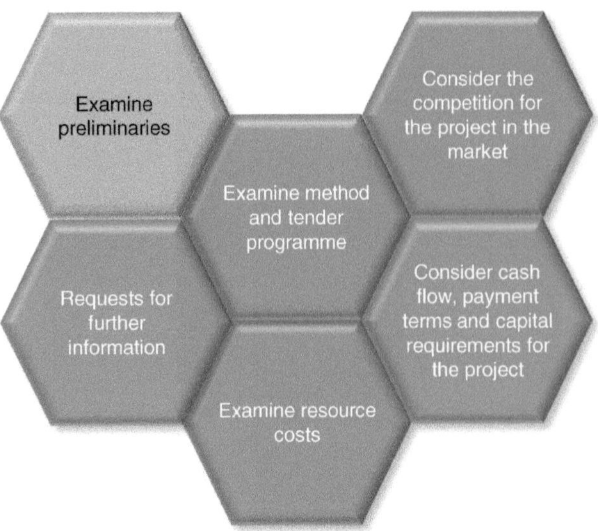

The preliminaries constitute a major part of the project cost. They are complex, involving the consideration of the needs of any specialty/sub-contractors, such as site works, accommodation and their impact on the schedule. The Preliminaries honeycomb shows the many interconnecting factors (hexagons) that need to be taken into consideration; each of those developing into honeycombs of their own. This complexity across tasks, people and processes requires careful consideration once a price has been estimated. An error at this stage could be costly. Understanding how the price has been reached is as important as the price itself.

6.2 Requests for further information

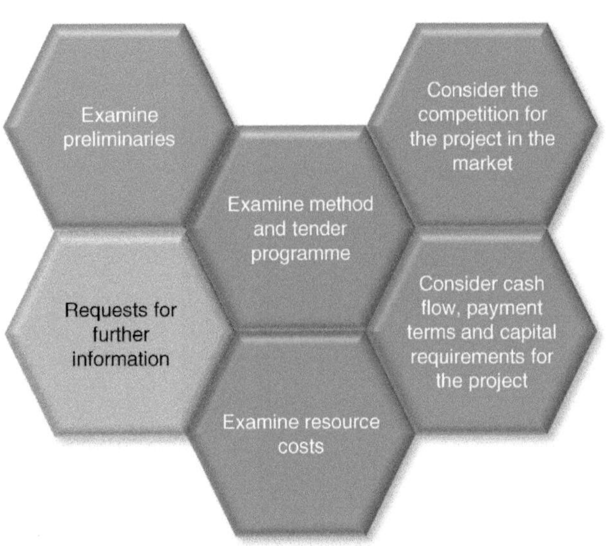

Requests for further information may come from the bid manager seeking clarification from a sub-contractor or the sub-contractor asking for more specific information in order to provide a reasonably accurate quote. Any requests need to be dealt with as quickly as possible to not overrun the period set aside for seeking sub-contractor quotes.

6.3 Examine method and tender programme

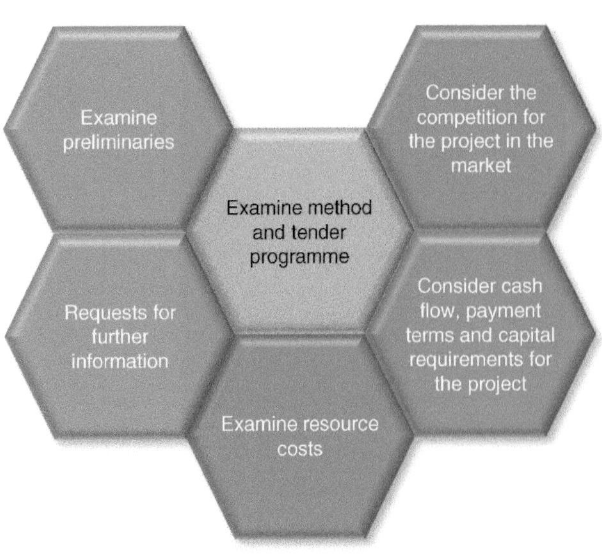

The method and tender programme should continue to be re-visited and updated where necessary. Using tender/bid software enables the bid team to be kept up-to-date with progress and any changes that have been made – part of the bid management process.

6.4 Examine resource costs

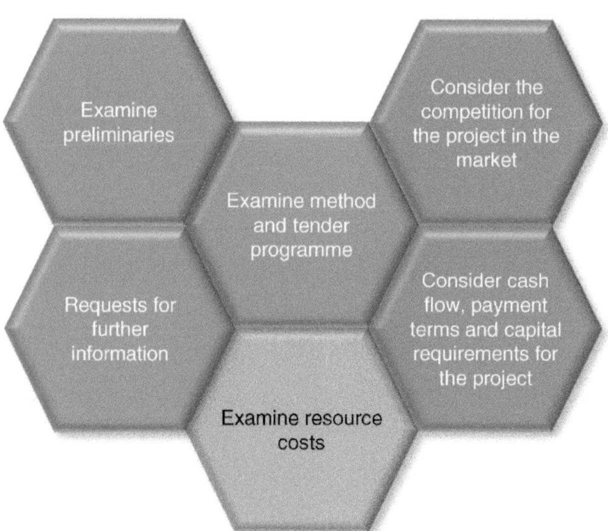

Resources such as labour, materials, plant and equipment may be priced individually within each task. There may be some sharing of costs with concurrent or successive tasks such as hire charges and labour. It is important to examine the resource costs to identify any duplication or where savings may be made by taking a more holistic view of the resource costs across the project. Any duplication between what is provided on site and what the specialty/subcontractor may need/provide should be considered. All resource costs have a time element, and so production planning is an important task to undertake.

6.5 Consider the competition for the project in the market

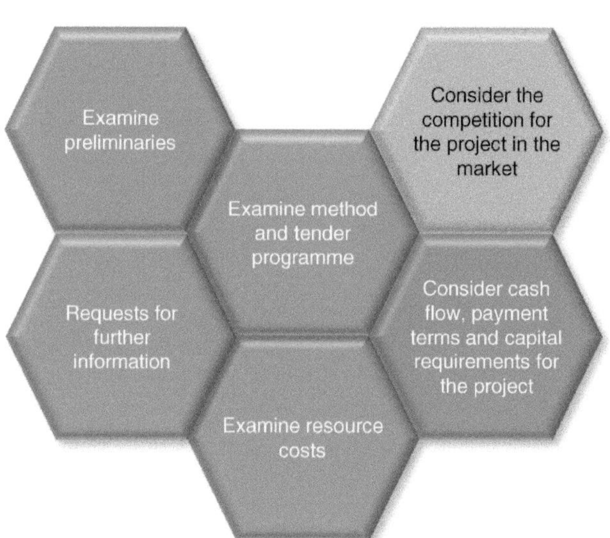

The section on Consider the likely competition looked at the factors in assessing the competitors likely to bid for the project being procured. It is also important to look at the demand for a particular project in the marketplace.

6.6 Consider cash flow and capital requirements for the project

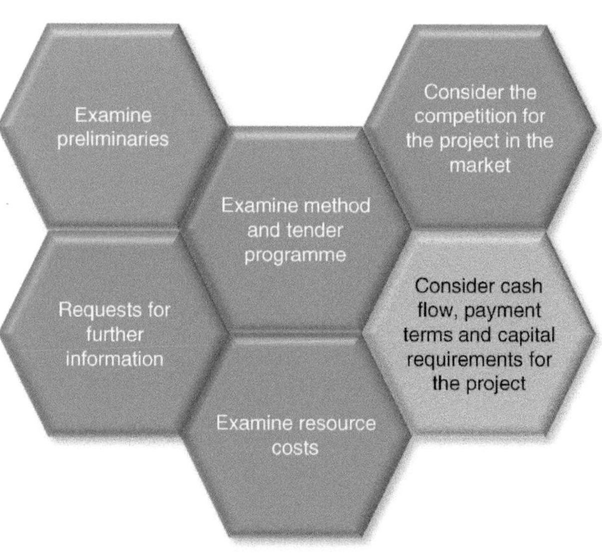

Cash and capital are major resources of the business. Cash flow is particularly important at the bid stage as it will need to be sufficient to fund the start of the work before the first payment is received. The cash flow for the whole business needs to be examined to avoid any impact on other ongoing projects. Capital may be required for any initial outlay on plant and equipment.

7 Bid assembly and adjudication

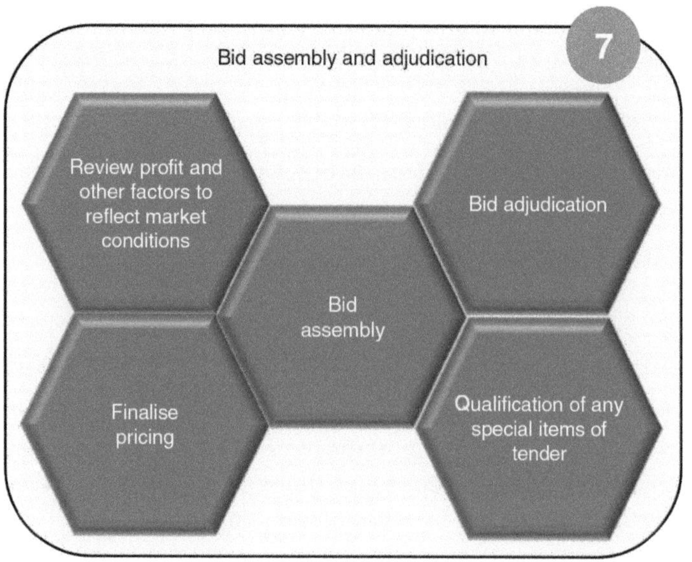

New Code of Estimating Practice, First Edition. The Chartered Institute of Building.

7.1 Finalise the pricing

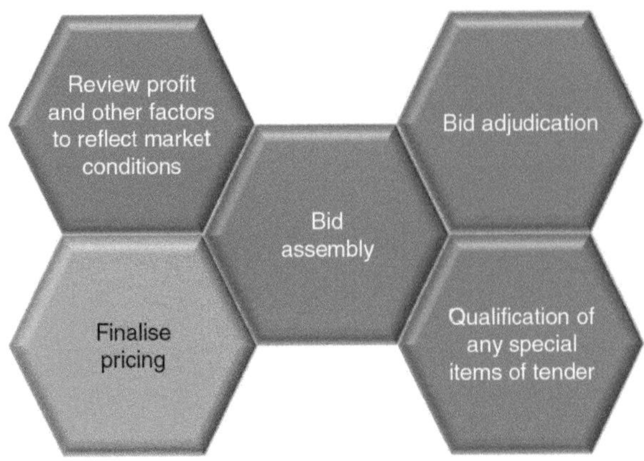

Pricing the bill of quantities (BoQ) needs to be finalised at this stage of the bid, detailing the assumptions made, the pricing information available and the underlying decisions. This provides information for the final review by the bid team.

It is important that the build-up of the rates for labour, materials, plant, equipment and specialty contractors can be traced and the processes documented. This supports any checking process that will be undertaken. By questioning during the review meeting, an estimator will explain how the principal rates were calculated.

The preparation of summaries of resources enables the estimator to list the most significant items, while management can assess the discounts, wastage factors and quantities allowed in the estimate. It is important that this reconciliation is made without relying entirely on the reports generated by a computer system.

Quotation totals from suppliers and specialty contractors can be checked against totals incorporated in the estimate.

For tenders based on drawings and specifications, an independent check on the principal quantities must be carried out in order to reduce the risk of a mistake, either leading to the winning of an undervalued bid or producing a high tender which may influence a client when giving opportunities to bid for future projects.

The estimator should prepare summaries of the resources making up an estimate so that management can assess the sources of labour, plant, materials and services; their price levels; price comparisons between suppliers and possible discounts available. The summaries are important for a number of reasons:

- The BoQ totals can be reconciled against the resource schedules
- They give management the ability to consider which area need to be focused on at the final review stage
- Adjustments to resource costs can be made where necessary.

Price fluctuations

In times of relatively low and predictable inflation, contractors are expected to submit tender prices that remain fixed for the anticipated duration of the work. Fluctuations will be paid under a fixed priced contract in certain circumstances, for example when materials are purchased during a period of compensable prolongation of the contract. For these purposes, the basic costs of materials are recorded in a list in an appendix to the bill of quantities.

Where the contract is to be fully fluctuating, the method of establishing the increases or decreases in cost need to be checked. It is common for a formula method to be adopted, and the rules for the application of the formula need to be studied carefully.

Where a firm price is to be submitted, an assessment must be made of the likely variations in cost during the proposed contract period. The tender works programme is an important tool in assessing the likely impact of price rises on certain elements of the project. There are some parts which will not need consideration such as provisional and prime cost (PC) sums and firm price quotations which fully comply with the conditions of contract.

Risk

Estimators generally make good risk managers as they develop a detailed understanding of the project as part of the estimating process and have the necessary skills of analytical assessment and quantification to complete and implement the risk register at tender stage.

Alternative tenders

The estimator should note in which part of the tender prepared may not meet the client's needs or where alternative approaches could give better value for money. It is acceptable to submit a compliant tender and to submit alternative tenders based on, for example shorter construction periods or sectional completions for further consideration by the client.

7.2 Bid adjudication/final review

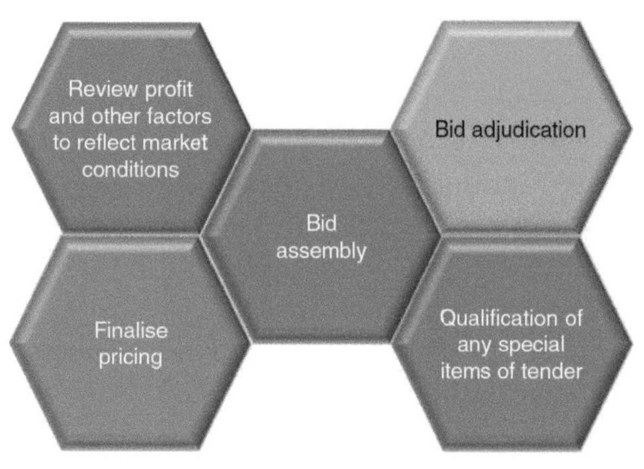

Reports, based on the cost-estimating processes undertaken, are presented at the final review meeting, also called the bid adjudication stage – see **Prepare Estimator's Report**. Summary reports and cost estimate breakdown, schedules and other bid-related information are considered by senior management.

Items to be considered include:

- The site
- The parties involved, for example client and consultants
- The contract
- The estimate
- The programme
- Project overheads.

An evaluation of alternatives, scope to improve profitability and risks which may be encountered should be considered at each stage of the estimating process and costs assessed in preparation for the settlement meeting. Clearly, in a strong competitive market, contractors need to find every opportunity to use products and processes which will enable the contract to support sufficient overheads and profits to maintain the company's objectives and satisfy the client by completing on time with the required specification.

When management considers risks at the meeting, an assessment is made of whether the priced documents comply with the tender requirements. The commercial and technical matters are checked to ensure, for example that the construction method, sequence and timing and any assumptions are acceptable, and it is assumed that the estimate has been correctly calculated with very few errors. In order to eliminate significant calculation errors, an estimating department needs procedures for ensuring that standards are maintained.

Finally, the mark-up will be added taking into account cash flow, scope, risk, head office overheads, profit, discounts and VAT. The mark-up will be applied to all costs above this level. It can be assumed that provisional sums and daywork will produce their own contribution to overheads and profit. These latter items may be considered as part of the project opportunity. If there are any late quotes or increased costs, these will be dealt with at this stage along with a decision on the process to deal with these.

At the meeting there may be other decisions to be made before proceeding to the submission stage, including:

- What documents will be submitted with the tender if nothing is specified in the invitation to tender? Some contractors may wish to prepare additional information such as a tender stage construction programme, company profile and printed brochures (if not previously provided during pre-qualification), whereas others reserve their efforts, and ideas, until they know that their tenders are under serious consideration.
- If an alternative method (or design) is to be offered, then another price may need to be settled by the review panel.
- Qualifications, which vary the requirements of the tender documents, are generally not permitted by clients. On the other hand, there may be circumstances where they cannot be avoided, and management must decide upon a suitable approach.
- In order to provide a positive cash flow for the duration of the project, it is important to decide how rates are to be apportioned in a bill of quantities to be submitted at the tender stage. Since any artificial alteration of prices will bring additional risks as well as opportunities, it is for management to agree a strategy.
- Evaluation of the risks that have arisen during the estimating process. For example, has there been a change of personnel involved in the preparation of the estimate?

For example:
'During the tender period, we identified some savings which can be brought about by small technical changes ...'
'We would need to clarify some of the contractual matters before entering into an agreement, but do not expect this to affect our price ...'.
Brook (2008, p. 244)

Where the bid is a two-stage process, that is where the selection of the contractor is made on limited criteria such as preliminaries and a percentage for overheads and profit, the final tender total may not be established until completion of the second stage process. It is important that a further review be carried out at the end of this process to ensure that the targets and margins set at the initial tender meeting have been maintained through the detailed pricing stage and that the completed tender still meets management expectation.

Bid adjudication decisions are highly subjective and are influenced by personal opinions and biases and based on experience, intuition and risk attitude. There is a level of inconsistency between each bid as the bid management team may be different for each project, and historical information may not always be reliable in a rapidly changing market.

7.3 Qualification of any special items of tender

Qualification of any special items may include the transfer of risk. Other qualifications may arise when:

- The wording of a performance bond is not acceptable.
- The duration suggested is too short.

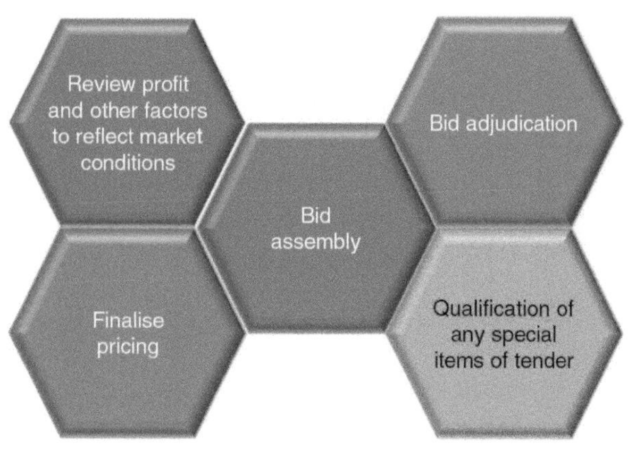

- There are late amendments, with little time to fund/organise extra responsibilities.
- There is a change to the named specialty contractors because one of them has defaulted.

The qualifications may not be detailed but written in general terms.

In some case, an alternative tender may be put forward. This could show an alternative offer to reduce the construction cost and/or shorten the duration. An alternative bid may propose a different way of sharing the risks.

8 Pre-production planning and processes

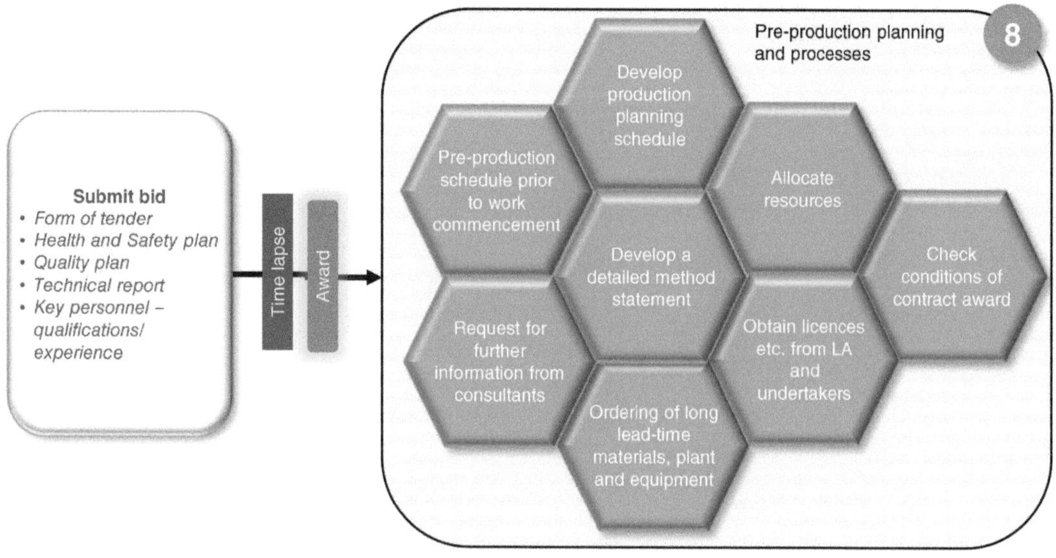

Pre-production planning and processes

8.1 Develop a detailed method statement

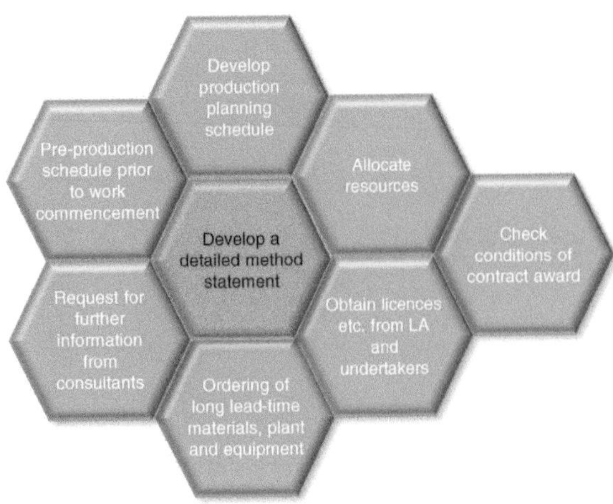

The method statement at the pre-tender stage would have had insufficient detail (unless it was a major project) because of the timescale of the bid process and the amount of information available. Once the bid is successful, the contractor needs to ensure that there is a detailed method statement. The method statement provides vital information to the estimator – see ***Method statement*** in the ***Principles and Processes*** sections and ***Operational estimating*** in the ***Principles*** section.

The method statement is much more detailed than a production planning schedule. It would typically include:

- Scope of work
- Location, work limits, site access and so on
- Known hazards identified in the risk assessment
- Landlord/site owner arrangements
- Access arrangements, any 'permits to work' and so on
- Responsible person(s)
- Briefing/communication arrangements
- Monitoring
- Operational sequence – how the work be structured and organised to be carried out in a safe manner
- How the controls detailed in the risk assessment will be implemented.
- Personnel safety arrangements – personal protective equipment (PPE) required and procedure for safe working
- Labour
- Plant/equipment (including safe working practices and checks)
- Materials (include potential safety issues such as manual handling and storage and disposal)
- Deliveries – identify routes and drop locations
- Emergency arrangements – identify first aid and fire or other emergency procedures, first aiders and location of first aid and fire equipment
- Environmental management – detail controls of harmful emissions to air, water and land.

Production planning is the process of organising and developing a plan of daily actions to be executed to complete a production process.

8.2 Develop production planning schedule

A production planning schedule should include activities; activity description; activity; duration; start and finish times/dates and the resources required.

The schedule mirrors the method statement in terms of activities but does not detail the way in which the activity should be undertaken. A bar or Gantt chart is also developed, which will highlight overlapping (concurrent work), clashes and peak resource periods. The increasing use of software in schedule planning allows the user to integrate easily processes across production and communicate the schedule to stakeholders. There is a wide variety of software designed to support the development of a schedule.

Scheduling is the process of building a time-based graphical representation of the desired goals describing task duration, resources, time constraints

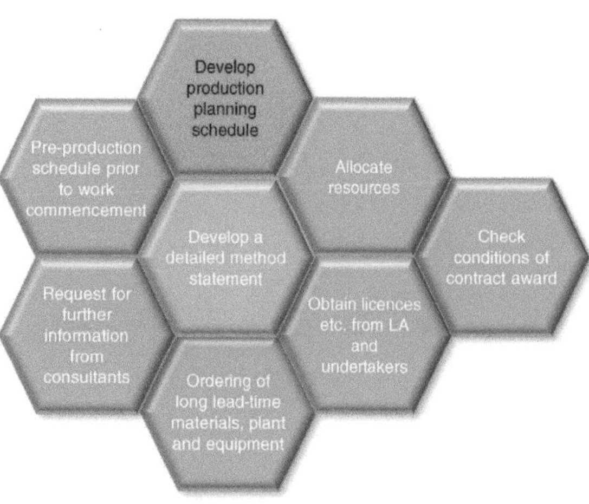

The schedule is used by a wide range of people in the production phase, and so needs to be accessible, up-to-date and, above all, realistic. The production planning schedule should be 'owned', that is its maintenance and updating managed by a specified person/people.

Production planning should include the coordination of trade contractors, planning of the material supply chain, continuous availability of work and contingencies for possible uncertainties involved in completing a task.

Developing and maintaining a production planning schedule is an iterative process that should constantly seek value in the use of resources. It needs to be dynamic to ensure optimum efficiency and productivity.

8.3 Allocate resources

Resources may be either constrained or unconstrained. This distinction is important in the allocation of resources and the planning involved in making sure that the right resource is available at the right time.

A resource is an entity that contributes to the execution of project activities such as manpower, material, money, equipment, time or space.

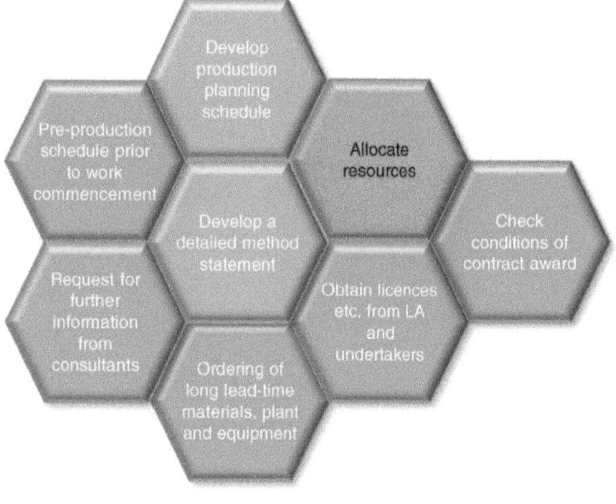

Time is an important factor as waiting for resources can cost money by causing delays in not only the activity requiring the delayed resource but also associated concurrent or following activities – see ***Ordering of long lead-time materials, plant and equipment***.

Levelling is a technique in which start and finish dates are adjusted based on resource constraints with the goal of balancing demand for resources with the available supply.

The allocation of resources is not just about their availability, it also concerns their quantities, suppliers, deliveries and storage. Resource levelling is important, that is, managing the peaks and troughs in the demand and supply of resources.

Resource smoothing differs from resource levelling in that it occurs when the time constraint takes priority. In the case of levelling, the priority is the limitations on resource availability. Both practices have a direct impact on the project duration.

8.4 Obtain licences and so on from local authority and utilities organisation

A number of licences may be required for the project and, as the application for these is often very time-consuming, some forward planning is required.

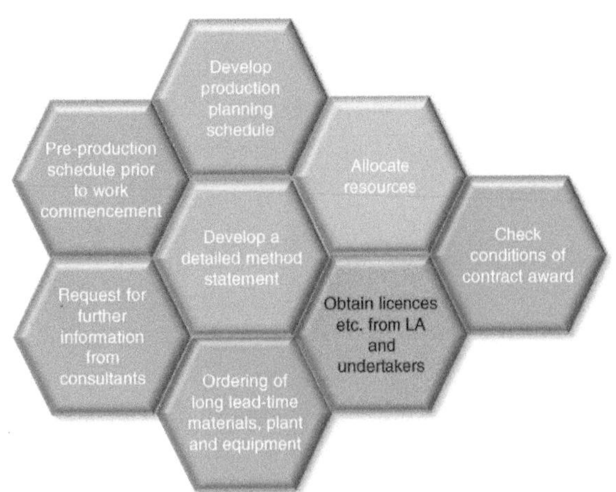

Permissions/licences from the local authority may include:

- Reserved matters, for example the detail of materials, design and access or anything not agreed at the initial planning stage
- Advertising
- Parking restrictions
- Highway matters
- The erection of scaffolding hoarding on the public highway
- Placing a skip on the public highway
- Temporarily placing other building materials on the public highway
- Placing a projection over the public highway
- Excavating in the public highway
- Oversailing (e.g. where a crane jib crosses over a public highway/footway)
- Wildlife survey
- Hazardous substances (e.g. the storage of chemicals and gases or the extraction of groundwater from an artesian well)
- Scaffolding
- Hoarding.

Statutory undertakers, such as utility companies, require licences, wayleaves and so on. There is often long lead periods required by these providers, and they should be informed as soon as possible in respect of any diversion or increase in existing supplies. Connections to gas, electricity, water and sewerage, including temporary services, need to be arranged well in advance.

8.5 Ordering of long lead-time materials, plant and equipment

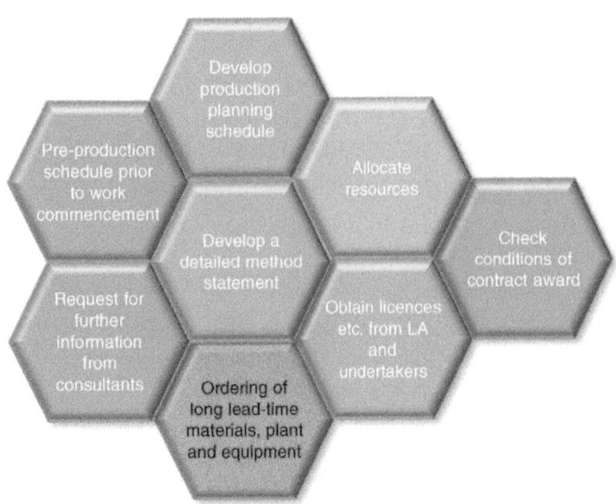

Some materials, plant or equipment are not available at short notice. Those with a long lead-in time need to be identified, and the process of ordering them begun as soon as possible in order to avoid delays which will impact the project's critical path. Lead times may also be affected by variables such as strikes, holiday periods, inclement weather and so on. Items coming from abroad that require transportation may take longer to arrive, and holiday periods and market conditions in other countries may also be different. Tables giving indications of lead times are available, for example Building magazine's quarterly 'lead times'.

8.6 Request for further information from consultants and specialty contractors

The short time span of the bid process means that some detailed information is not sought until the bid has been won. This is not only due to time constraints but also budgetary ones. Once successful, the contractor's estimator needs to request further information where necessary.

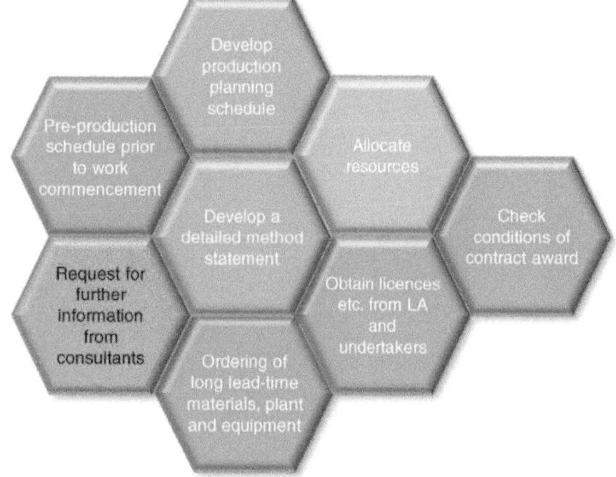

8.7 Pre-production schedule prior to work commencement

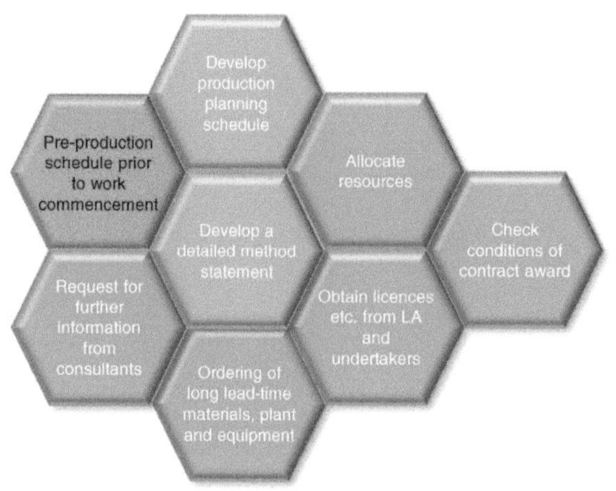

CDM (2015) requires the client to make suitable arrangements for the management of the project to ensure that the project is undertaken so that health and safety risks are managed (HSE, 2015b). Where there is more than one contractor, a principal designer and principal contractor need to be appointed. Pre-construction information must be provided from the feasibility stage by the client and:

- be relevant to the particular project;
- have an appropriate level of detail and
- be proportionate to the risks involved.

This information enables a pre-production schedule to be created.

8.8 Check conditions of contract award

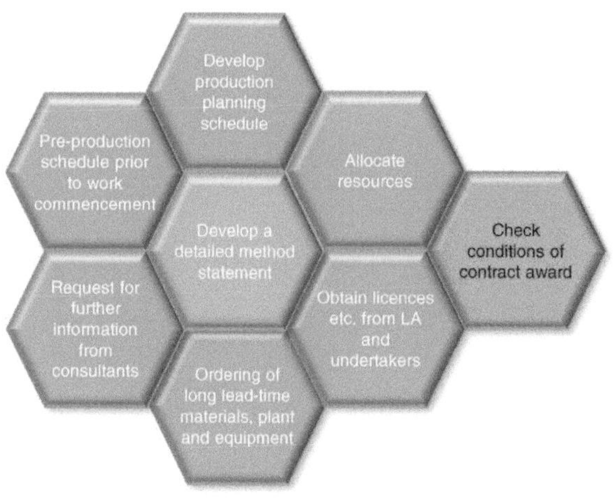

The contract terms and conditions should be reviewed following the award of the contract, comparing it with the contract that accompanied the original tender. Any differences should be noted, and the rights and obligations of each party fully understood.

It may be necessary to update the project/programme plan with the involvement of both parties, to reflect the actual date of effectiveness as well as milestones/deliverables of the contract and any changes which may have occurred since it was planned.

If there is a contract manager, he/she will document contract performance for the following reasons as this can constitute proof of performance and provide evidence in the event of a dispute. Any documentation will be used for audit purposes.

9 Site production

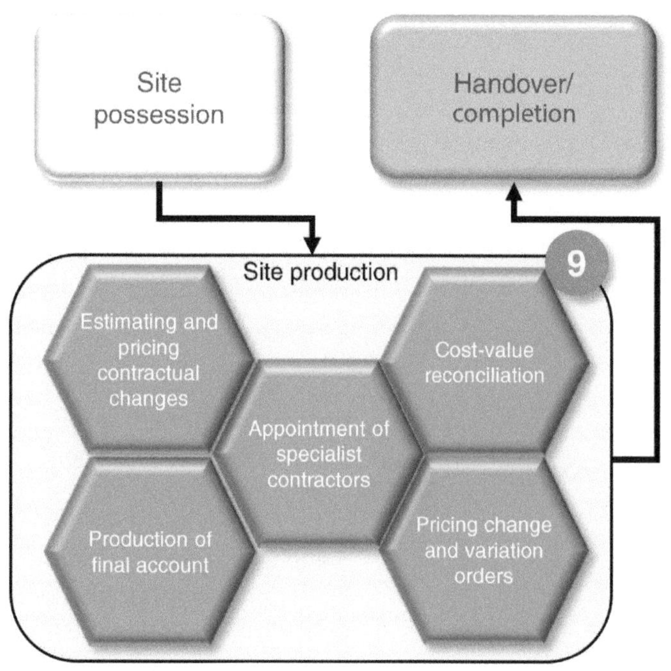

9.1 Pricing change and variation orders

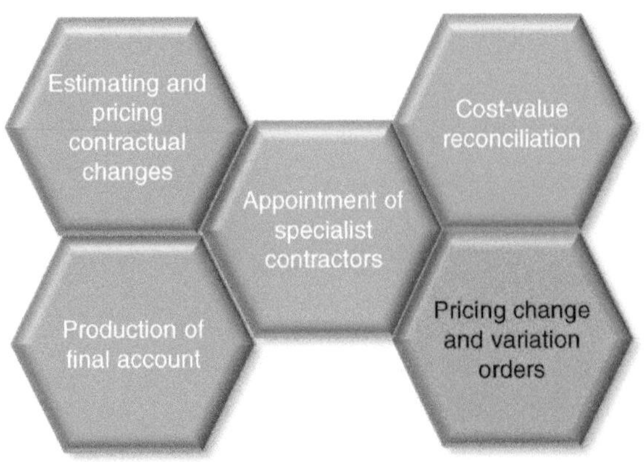

It is commonplace for there to be changes to the original contract following a successful bid. There may be changes in the scope (size, quality, etc.), unforeseen conditions or professional errors and omissions. If the design is not complete at the bid stage, the likelihood of the need for a variation order* is high. Any change necessitates the re-calculation of the project price and consideration of the likely impact on related/dependent work packages and the project duration.

A change order is work that is added to or deleted from the original scope of work of a contract, which alters the original contract amount and/or completion date.

Change orders describe the work that needs to be added to (or deleted from) the original agreement. They can significantly affect both the time and cost of the project and are agreed by the client contractor and designer. Change orders are a common feature of construction projects and can be the result of a variation in the original design to achieve efficiencies; inaccurate original estimates; inefficiencies in the production process which then require additional time, money and resources and changes required by the client. A change order has to be generated to describe the new work needed and the price to be paid. Once approved, it becomes an alteration to and so part of the original contract.

9.2 Appointment of specialist contractors

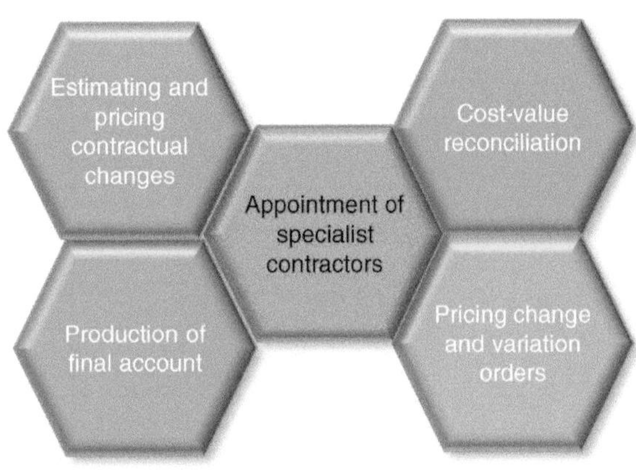

Once the bid has been won, the specialist contractors who had tendered for various work packages need to be confirmed and their costs have been finally agreed. Changes in the original tender documentation/requirements may mean that specialty contractors' costs may be higher or lower than the original estimate. More detailed information post-bid allows the estimator to refine estimates.

9.3 Estimating and pricing contractual claims

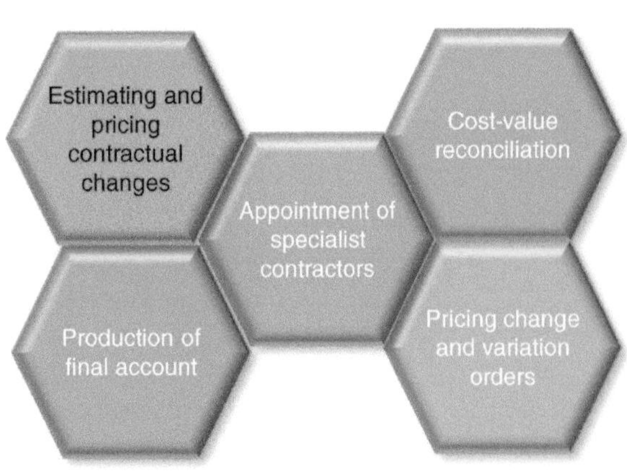

Claims by the contractor/consultant may arise from a contract for a number of reasons:

- Delays or changes in the works.
- Non-payment of fees or breach of copyright.

On the other hand, the client may claim against the contractor for delays, defects, design failure or lack of appropriate supervision.

* Variation order is a term used in the UK, while the USA use the term 'Change Order'. FIDIC has adopted the term 'Variation'. All mean the same thing.

Estimating and pricing any contractual claims is an important task as it impacts the final outturn of the cost of the project. The estimate should consider a number of issues:

- *Actual costs*: rather than what was included in the bid document. These can include allowance for inflation due to any delays
- *Preliminaries*: these already include set-up, running and dismantling costs, therefore, only running costs at the point of breach (and any delay caused) should be calculated
- *Disruption*: this can cause loss of productivity, a difficult concept to estimate. A comparison of productivity before the disruption and after could be a good guide
- *Overheads*: Hudson's formula is widely accepted by the courts and is:

Certain costs would need to be subtracted from this formula: Credit for visiting supervisors' time; any additional overhead recovered within the final account, for example variation account; where resources were re-deployed due to any delay.

$$\frac{\text{HQ profit \%}}{100} \times \frac{\text{Contract sum}}{\text{Contract period (weeks)}} \times \text{Delay (weeks)}$$

- *Loss of profit/opportunity costs*: this is only appropriate if it can be proved that the contractor was prevented from making profit elsewhere. Any profit made due to extra work and priced within the final account should be subtracted
- *Finance charges and interest*: these are the recoverable amounts based on any extra capital required to fund the costs of any breaches. The interest rates need to be proven and be reasonable and, if financed through the corporate business, the rate equates to that received from monies placed on deposit.

9.4 Production of final account

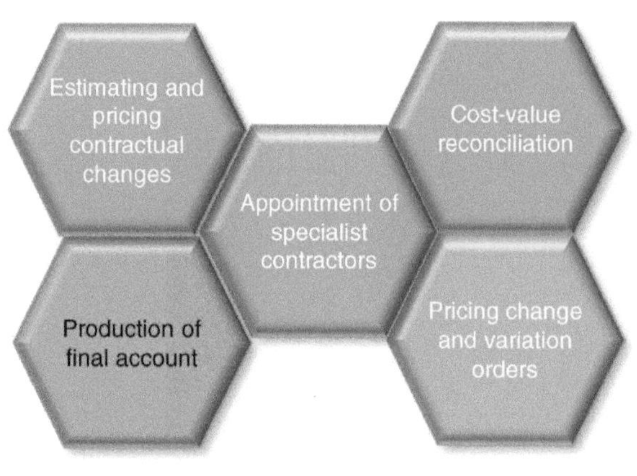

The final account brings together all the financial aspects of the project within the contract, taking account of any adjustments that were needed. It should be a fair and reasonable valuation of costs, to be submitted to the contract administrator for the completion certificate to be issued. Reaching agreement may involve some negotiation. The sum according to the contract may need to be adjusted due to:

- Variations
- Fluctuations
- Prime cost sums
- Provisional sums
- Payments to nominated sub-contractors or nominated suppliers
- Statutory fees
- Payments relating to the opening-up and testing of the works
- Loss and expense
- Liquidated and ascertained damages
- Contra claims imposed as a result of the contractor's operations (such as a third-party claim resulting from contractor negligence or contractual breach, for example, flooding a neighbour's property)
- The release of any remaining retention.

The amount of time it will take to produce the final account should not be underestimated; it is a complex process. The process can be simplified by agreeing any adjustments to the contract sum as the project progresses. Agreement of a draft version by the quantity surveyors representing both the client and the contractor will help.

Payment of the final account is made at the end of the defects liability period and once any patent defects have been rectified.

9.5 Cost–value reconciliation

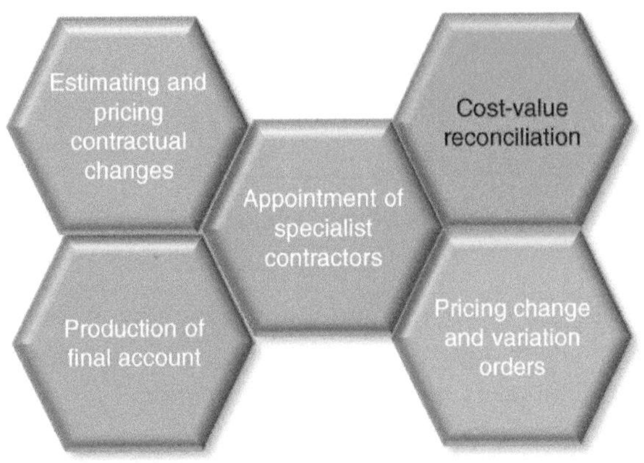

A cost–value reconciliation (CVR) is undertaken to monitor what is being spent against what was allowed for in the budgets. Measuring the actual costs against the value of the works plus profit provides a project balance sheet. CVRs can help measure profitability and keep the management team informed of the likely outturn. CVR can underpin financial reporting in interim valuations and also in final account negotiations.

CVR can help to identify budgetary problems at an early stage. However, the fragmented nature of construction projects, often with numerous specialty contractors, means that CVR is not an easy task. Whilst spreadsheets have been the most common way of keeping records of costs and value, they are often task-based and do not give a more holistic view. Software is available that will accomplish a more integrated approach.

Bibliography and References

AACE (2016) **Cost estimate classification system – as applied in engineering, procurement, and construction for the process industries**. International recommended practice 18R-97.

Abdou, A., Lewis, J. and Alzarooni, S. (2004) Modelling risk for construction cost estimating and forecasting: a review. *In:* Khosrowshahi, F. (Ed.), *Proceedings 20th Annual ARCOM Conference*, 1–3 September 2004, Edinburgh, UK. Association of Researchers in Construction Management, Vol. 1, pp. 141–52.

Akintoye, A. and Fitzgerald, E. (2000) A survey of current cost estimating practices in the UK, Construction Management and Economics, Vol. 18, Issue 2, pp. 161–172.

Al-Hasan, M., Ross, A. and Kirkham, R. (2005) An investigation into current cost estimating practice of specialist trade contractors. *3rd International Built and Human Environment Research Week*, pp. 566–575.

American Institute of Architects (2007) Integrated project delivery: a guide. American Institute of Architects, Washington DC. pp. 57.

Baccarini, D. (2004) Estimating project cost contingency - a model and exploration of research questions. *In:* Khosrowshahi, F. (Ed.), *20th Annual ARCOM Conference*, 1–3 September 2004, Heriot Watt University. Association of Researchers in Construction Management, Vol. 1, pp. 105–13.

Baldwin, A. and Bordoli, D. (2014) **Handbook for Construction Planning and Scheduling**. Wiley, 433pp. ISBN: 9780470670323.

Banaitiene, N. and Banaitis, A. (2012) Risk management in construction projects. Chapter 19 *In:* Banaitiene, N. (Ed.), Risk Management - Current Issues and Challenges. ISBN: 978-953-51-0747-7.

Bernstein, P.L. (1998) **Against the Gods: The Remarkable Story of Risk**. Wiley. 383p.

BIS (2013) **Supply chain analysis into the construction industry report for the construction industrial strategy**. BIS Research Paper No. 145. London: Department for Business Innovation and Skills, 127pp.

Brook, M. (2008) **Estimating and Tendering for Construction Work**. Elsevier/Butterworth-Heinemann, 359pp.

BSI (1989) **Workmanship on building sites - Part 1: Code of practice for excavation and filling**. BS 8000–1:1989. BSI Standards Publication. ISBN: 0 580 17660.

BSI (1995) **Glass for glazing**. BS 952-1-1995. BSI Standards Publication.

BSI (2003) **Temporary works equipment – Part 1: Scaffolds – performance requirements and general design**. BS EN 12811-1-2003. BSI Standards Publication.

BSI (2004) **Falsework – performance requirements and general design**. BS EN 12812–20. BSI Standards Publication.

BSI (2008) **Falsework – performance requirements and general design**. BS EN 12812:2004. BSI Standards Publication. ISBN: 0 580 44834 7.

BSI (2009) **Code of practice for earthworks**. BS 6031:2009. BSI Standards Publication. ISBN: 978 0 580 72749 8.

BSI (2010) **Construction procurement-Part 1 processes, methods, and procedures**. BS ISO 10845-1-2010. BSI Standards Publication. ISBN: 978 0 580 63465 9.

BSI (2011) **Code of practice for temporary works procedures and the permissible stress design of falsework**. BS 5975:2008+A1:2011. BSI Standards Publication. ISBN: 978 0 580 74257 6.

BSI (2011) **Construction procurement-Part 2 Formatting and compilation of procurement documentation**. BS ISO 10845-2-2011. British Standards Publication. ISBN: 978 0 580 66104 4.

Bussell, M., Lazarus, D. and Ross, P. (2003) **Retention of Masonry Facades – Best Practice Site Handbook**. CIRIA. ISBN: 0 86017 589 8.

Carpenter, J. (2012) Temporary works and the structural engineer. The Structural Engineer, Vol. 90, Issue 12.

Carr, R.I. (1989) Cost estimating principles. Journal of Construction Engineering and Management, Vol. 115, Issue 4, pp. 545–551.

Chan, E.H.W. and Au, M.C.Y. (2008) Relationship between organizational sizes and contractors' risk pricing behaviors for weather risk under different project values and durations. Journal of Construction Engineering and Management, Vol. 134, Issue 9, pp. 673–680. ISSN: 0733-9364/2008/9-673–680.

New Code of Estimating Practice, First Edition. The Chartered Institute of Building.

Choon, T.T. and Ali, K.N. (2008) A review of potential areas of construction cost estimating and identification of research gaps. Jurnal Alam Bina, Jilid, Vol. 11, Issue 2, pp. 61–72.

Chua, D.K.H. and Li, D. (2000) Key factors in bid reasoning model. *Journal of Construction Engineering and Management*, Vol. 126, Issue 5, pp. 349–357.

CIOB (2011) **Guide to Good Practice in the Management of Time in Complex Projects**. Oxford: Wiley-Blackwell.

CIOB (2009) **Code of Estimating Practice, 7th Edition**. Ascot: Chartered Institute of Building, 157p.

CIPS (2007) **How to Develop a Waste Management and Disposal Strategy**. The Chartered Institute of Purchasing & Supply.

CIRIA (1996) **Control of Risk: A Guide to the Systematic Management of Risk From Construction**. London: Construction Industry Research and Information Association.

CIRIA (2003a) **Crane Stability on Site – An Introductory Guide**. CIRIA C703. London: Construction Industry Research and Information Association and Department of Trade and Industry.

CIRIA (2003b) **Retention of Masonry Facades – Best Practice Guide**. CIRIA C579. London: Construction Industry Research and Information Association.

CIRIA (2004) **Drainage of Development Sites – A Guide**. X108. Construction Industry Research and Information Association and HR Wallingford.

CIRIA (2009) **Unexploded Ordnance (UXO) a Guide for the Construction Industry**. London: CIRIA. ISBN: 978-0-86017-681-7.

CIRIA (2011) **Working with Wildlife: Guidance for the Construction Industry**. CIRIA C691. London: Construction Industry Research and Information Association.

Civil Engineering Dictionary (2015) http://www.aboutcivil.org/setting-out.html. *Accessed 12th August 2015.*

Concrete Society (2010) **Concrete on Site 3 – Formwork**. Camberley, Surrey: The Concrete Society.

Couzens, A., Skitmore, M.R., Thorpe, T. and McCaffer, R. (1996) A decision support system for construction contract bidding adjudication. Civil Engineering Systems, Vol. 13, pp. 121–139.

CPA (2008) **Maintenance, Inspection and Thorough Examination of Tower Cranes**. CPA Best Practice Guide. TCIG 0801. Construction Plant-Hire Association.

Croydon (2012) **A Practical Guide to Drafting a Construction Logistics Plan**. Croydon Council, 16pp.

DEFRA (2012) **Statutory guidance for blending, packing, loading, unloading and use of cement**. Process Guidance Note 3/01(12).

Dell'Isola, M.D. (2003) Detailed cost estimating. *In:* **The Architect's Handbook of Professional Practice**. American Institute of Architects, published by John Wiley & Sons, Inc.

Department for Communities and Local Government (2012) National Planning Policy Framework. pp. 65.

Department of Transport (2008) **Design Manual for Roads and Bridges**. http://www.standards forhighways.co.uk/ha/standards/dmrb/index.htm. *Accessed August 24th 2015.*

Designing Buildings Wiki (2015) **Designing Buildings** Wiki. http://www.designingbuildings.co.uk/wiki/Home. *Accessed 5th August 2015.*

Eldosouky, I.A., Ibrahim, A.H. and Mohammed, H.E.-D. (2014) Management of construction cost contingency covering upside and downside risks. Alexandria Engineering Journal, Vol. 53, Issue 4, pp. 863–881.

European Commission (2011) **Non-Binding Guide to Good Practice for Understanding and Implementing Directive 92/57/EEC**. Luxembourg: Publications Office of the European Union.

Fadiya, O.O. (2012) **Development of an integrated decision analysis framework for selecting ICT-based logistics systems in the construction industry**. Unpublished PhD Thesis. University of Wolverhampton.

Fang, S.Y. and Ng, T. (2011) Applying activity-based costing approach for construction logistics cost analysis. Construction Innovation, Vol. 11, Issue 3, pp. 259–281.

FTA (2016) Designing for deliveries. Freight Transport Association.

Glass and Glazing Federation (2005) **Code of practice – Glass handling, storage and transport**.

Greenhalgh, B. (2013) **Introduction to Estimating for Construction**. Abingdon, Oxon: Routledge, 216pp.

Hackett, J.A. (2010) **The role of the cost estimators in UK construction: A case for and steps towards an estimating profession**. Unpublished PhD Thesis. The University of Birmingham.

Hillson, D.A. and Hulett, D.T. (2004) Assessing risk probability: alternative approaches. *Proceedings of PMI Global Congress 2004 EMEA*, Prague, Czech Republic.

HM Government (2010a) **The Building Regulations 2010 – Site preparation and resistance to contaminants and moisture. Part C**. ISBN: 978-1-85946-509-7.

HM Government (2010b) **The Building Regulations 2010 – Drainage and waste disposal**. Part H. ISBN: 978-1-85946-599-8.

HM Treasury (2011) **The Green Book – Appraisal and Evaluation in Central Government**. London: TSO.

HSE (1992) **Façade Retention: Guidance Note GS 51**. Health and Safety Executive. http://regulations.completepicture.co.uk/pdf/Health%20and %20Safety/Facade%20retention.pdf. *Accessed 20th August 2015.*

HSE (1997) **Lighting at Work**. HSG38. Health and Safety Executive. ISBN: 978-0-7176-1232-1.

HSE (2005) **The Control of Noise at Work Regulations**. 130p. Contains public sector information published by the Health and Safety Executive and licensed under the Open Government Licence.

HSE (2006) **Health and Safety in Construction**. HSG150. Health and Safety Executive. http://www.hse. gov.uk/pubns/priced/hsg150.pdf. *Accessed 19th August 2015.*

HSE (2010) **The Management of Temporary Works in the Construction Industry**. Health and Safety Executive. SIM 02/2010/04. Online document http://www.hse.gov.uk/foi/internalops/sims/constrct/2_10_04.htm. *Accessed 18th August 2015.*

HSE (2011) **Workplace Exposure Limits**. EH40. Health and Safety Executive. ISBN: 978-0-7176-6446-7.

HSE (2013) **Control of Substances Hazardous to Health - The Control of Substances Hazardous to Health Regulations 2002 (as amended)**. Health and Safety Executive. ISBN: 978-0-7176-6582-2.

HSE (2015a) **Scaffolding Checklist**. http://www.hse.gov.uk/construction/safetytopics/scaffolding info.htm. *Accessed 19th August 2015.*

HSE (2015b) **Managing Health and Safety in Construction: Construction (Design and Management) Regulations 2015**. Health and Safety Executive, 90pp. *Contains public sector information published by the Health and Safety Executive and licensed under the Open Government Licence.* ISBN: 978-0 7176-6626-3.

HSE (2015c) **Safety Signs and Signals: The Health and Safety (Safety Signs and Signals) Regulations 1996**. Health and Safety Executive, 49pp. *Contains public sector information published by the Health and Safety Executive and licensed under the Open Government Licence.*

ICE (2014) **CESMM4: Civil Engineering Standard of Method and Measurement (CESMM4 Series), 4th Revised Edition**. ICE Publishing.

Illingworth, J.R. (2002) **Construction Methods and Planning**. CRC Press.

Investopedia (2015) **Contractors' All Risks (CAR) Insurance**. http://www.investopedia.com/terms/c/contractors-all-risks-car-insurance.asp. *Accessed 5th August 2015.*

JCT (2012) **Tendering 2012. 2012 Practice Note**. London: Sweet and Maxwell.

JCT (2017) **2017 Practice Note**. London: Sweet and Maxwell. ISBN 978-0-414-06447-8

Ji, S.-H., Park, M., Lee, H.-Y., Lee, H.-S., Ahn, J., Kim, N. and Son, B. (2011) Military facility cost estimation system using case-based reasoning in Korea. Journal of Computing in Civil Engineering, Vol. 25, Issue 3, pp. 218–231.

Joint Federal Government (2012) **Guide to Cost Predictability in Construction: An Analysis of Issues Affecting the Accuracy of Construction Cost Estimates**. Joint Federal Government and Industry Cost Predictability Taskforce. Canadian Construction Association.

Kim, H.-J., Seo, Y.C. and Hyun, C.T. (2012) A hybrid conceptual cost estimating model for large building projects. Automation in Construction, Vol. 25, pp. 72–81.

Knutson, J. (Ed.) (2001) **Project Management for Business Professionals: A Comprehensive Guide**. Hoboken, NJ: John Wiley & Sons, Inc., 626pp.

Kundera, M. (2000) **Immortality**. Faber & Faber. pp. 400.

Laryea, S. and Hughes, W. (2011) Risk and price in the bidding process of contractors. Journal of Construction Management and Engineering, Vol. 137, Issue 4, pp. 248–258.

Long, D., Nguyen, L.D., Kneppers, J., de Soto, B.G. and Ibbs, W. (2010) Analysis of adverse weather for excusable delays. Journal of Construction Engineering and Management, Vol. 136, Issue 12, pp. 1258–1267.

Lou, E.C.W. and Goulding, J.S. (2008) Building and construction classification systems, Architectural Engineering and Design Management, Vol. 4, Issue 3–4, pp. 206–220.

Madhavi, T.P., Varghese, S. and Sasidharan, R. (2013) Material management in construction – a case study. *International Journal of Research in Engineering and Technology*. IC-RICE Conference issue.

March, C. (2009) **Finance and Control for Construction**. London: Spon Press.

Moselhi, O., Gong, D. and El-Rayes, K. (1997) Estimating weather impact on the duration of construction activities. Canadian Journal of Civil Engineering, Vol. 24, Issue 3, pp. 359–366.

Mossman, A. (2008) More than materials: managing what's needed to create value in construction. *2nd European Conference on Construction Logistics*, ECCI, Dortmund.

NCHRP (2014) **Climate Change, Extreme Weather Events, and the Highway System: Practitioner's Guide and Research Report 750**, Volume 2. National Cooperative Highway Research Program.

Nevada DOT (2012) **Risk Management and Risk-Based Cost Estimation Guidelines**. Nevada Department of Transport.

NSCC (2007) **Reduce, Reuse, Recycle – Manage Your Waste**. National Specialist Contractors Council.

Olatunji, O.A., Sher, W. and Ogunsemi, D.R. (2010) The impact of building information modelling on construction cost estimating. Proceedings of the *W055 - Special Track 18th CIB World Building Congress* May 2010 Salford, UK.

Oo, B.L., Ling, F.Y.Y. and Soo, A. (2014) Information feedback and bidders' competitiveness in construction bidding. Engineering, Construction and Architectural Management, Vol. 21, Issue 5, pp. 571–585.

Picken, D.H. and Mak, S. (2001) Risk analysis in cost planning and its effect on efficiency in capital cost budgeting. Logistics Information Management, Vol. 14 Issue 5/6, pp. 318–329.

Picken, D.H., Mak, S. and Eng, C.S. (1999) An analysis of risk management in cost planning and its effect on efficiency in capital cost budgeting. *CIB W55 & W65 Joint Triennial Symposium Customer Satisfaction: A Focus for Research & Practice*, Cape Town, 5–10 September 1999.

PMBook (2015) **Fundamental Scheduling Procedures**. http://pmbook.ce.cmu.edu/10_Fundamental_ Scheduling_ Procedures.html. *Accessed 17th August 2015.*

PPP Canada (2014) **Schematic Design Estimate Guide**. PPP Canada.

RICS (2010) **e-Tendering**. RICS Practice Standards, UK. ISBN: 978 1 84219 563 5.

RICS (2012a) **NRM 1: Order of Cost Estimating and Cost Planning for Capital Building Works, 2nd Edition**. Royal Institution of Chartered Surveyors.

RICS (2012b) **NRM 2: Detailed Measurement for Building Works**. Coventry, Royal Institution of Chartered Surveyors. ISBN: 978 1 84219 716 5.

RICS (2015a) **Bonds, Guarantees, Warranties and Third Party Rights**. https://www.isurv.com/site/scripts/documents.aspx?categoryID=383. *Accessed 5th August 2015.*

RICS (2015b) **Management of Risk, 1st Edition**, RICS Guidance Note. Royal Institution of Chartered Surveyors.

Rider Levett Bucknall (2015) **Riders Digest 2015: United Kingdom**. Rider Levett Bucknall.

Sabol, L. (2008) **Challenges in Cost Estimating with Building Information Modelling**. Washington, DC: Design and Construction Strategies, LLC.

Sadgrove, B.M. (2007) **Setting-Out Procedures for the Modern Built Environment**. C709. London: CIRIA.

Shash, A. (1993). Factors considered in tendering decisions by top UK contractors. Construction Management and Economics, Vol. 11, Issue 2, pp. 111–118.

Sinclair, N. Artin, P. and Mulford, S. (2002) Construction cost data workbook. *Conference on the International Comparison Program*, 11–14 March, 2002, Washington, DC, World Bank.

SJG Temporary Works Ltd (2015) **Presentation to Birmingham Health, Safety and Environment Association**. http://www.bhsea.org.uk/shad20145.pdf. *Accessed 18th August 2015.*

Smith, A.J. (1995) **Estimating, Tendering and Bidding for Construction**. Houndmills, Basingstoke: Macmillan Press Ltd, 256pp.

Smith, G.R. and Hancher, D.E. (1989) Estimating precipitation impacts for scheduling. Journal of Construction Engineering and Management, Vol. 115, Issue 4, pp. 552–566.

Solomon, G. (1993) **Cost Analysis, Preliminaries, The Marketplace Effect**. Chartered Quantity Surveyors. October, 9–11.

Staub-French, S. and Fischer, M. (2000) **Practical and research issues in using industry foundation classes for construction cost estimating**. CIFE Working Paper No. 56. Center for integrated facility Engineering, Stanford University.

Taboada, J.A. and Garrido-Lecca, A. (2014) Case study on the use of BIM at the bidding stage of a building project. *Proceedings IGLC-22*, June 2014 | Oslo, Norway.

Transportation Research Board (2014) **Strategic issues facing transportation, volume 2: climate change, extreme weather events, and the highway system**: Practitioner's Guide and Research Report. NCHRP Report 750. The National Academies of Sciences, Engineering, and Medicine.

Transport for London (2017) **Construction Logistics Plan Guidance**. 50p.

Trevor Sadd Associates (2005) **Preliminaries (general site costs) – A Guide to the Pricing and Use of Preliminaries in the Formulation of Budgets, Quotations and Tenders**. Trevor Sadd Associates Ltd.

TSO (2012) **Preliminaries (general site costs) – A Guide to the Pricing and Use of Preliminaries in the Formulation of Budgets, Quotations and Tenders**. Trevor Sadd Associates Ltd.

TSO (2012) Traffic Management Act 2004. The Stationery Office. *Contains public sector information published and is licensed under the Open Government Licence.*

TSO (2015) **The Public Contracts Regulations 2015**. The Stationery Office. *Contains public sector information published and is licensed under the Open Government Licence.*

TW Forum (2014) **The use of European standards for temporary works design - discussion document** TW/14/030. Temporary Works Forum. www.twforum.org.uk/media/47604/tw14.030__en_pt1_issued_july14.pd. *Accessed 24th August 2015.*

US Department of Energy (2011a) **Cost Estimating Guide**. DOE G 413.3-21.

US Department of Defense (2011b) **Handbook: Construction Cost Estimating**. United Facilities Criteria.

Vidalakis, C., Tookey, J.E. and Sommerville, J. (2011) The logistics of construction supply chains: the builders' merchant perspective. Engineering, Construction and Architectural Management, Vol. 18, Issue 1, pp. 66–81.

Water UK (2015) **Best practice guidelines for the production of estimates, quotations and terms for the provision of water mains on new developments**, 5pp.

Wielebski, S. (2013) **CIOB Carbon Action 2050 - Effective Management Of Construction Waste**. Chartered Institute of Building.

Wilson James (2015) **Construction Consolidation Centre**. www.wilsonjames.co. uk/case-study-6-construction-consolidation-centre-.html. *Accessed 4th September 2015.*

WRAP (2007) **Material Logistics Plan - Good Practice Guidance**. Oxford: WRAP.

Wu, S., Wood, G., Ginige, K. and Jong, S.W. (2014) A technical review of BIM-based cost estimating in UK quantity surveying practice, standards and tools. Journal of Information Technology in Construction (ITcon), Vol. 19, pp. 534–563.

Ying, E., Tookey, J. and Roberti, J. (2014) Addressing effective construction logistics through the lens of vehicle movements. Engineering, Construction and Architectural Management, Vol. 21 Issue 3, pp. 261–275.

Yoe, C. (2000) **Risk analysis framework for cost estimation**. IWR Report 00-R-9. US Army Corps of Engineers.

Zanen, P. and Hartmann, T. (2010) **The application of construction project management tools - an overview of tools for managing and controlling construction projects**. Working Paper #3. Center for Visualization and Simulation in Construction pp. 14.

Zhiliang, M. and Zhenhua, W. (2012) Framework for automatic construction cost estimation based on BIM and ontology technology. *Proceedings of the CIB W78 2012: 29th International Conference*, Beirut, Lebanon, 17–19 October.

Index